GREETINGS FROM EARTH

GREETINGS FROM EARTH

NEW AND
COLLECTED STORIES
BY
SCOTT BRADFIELD

Picador USA
New York

Some of these stories originally appeared in the United States as *Dream of the Wolf* by Alfred A. Knopf Inc. and in Great Britain as *The Secret Life of Houses* by Unwin Hyman Ltd.

The author gratefully acknowledges the following publications, in which some of the stories in this book originally appeared: *Ambit, Interzone, Omni, Other Edens II, Soho Square II, The Tampa Review, Tarot Tales, Twilight Zone Magazine, The New Gothic, The Los Angeles Times Magazine* and *Conjunctions.*

Grateful acknowledgement is made to CPP/Belwin, Inc. and International Music Publications for permission to reprint an excerpt from 'Duke's Place,' lyrics by Ruth Roberts, Bill Katz, and Robert Thiele and music by Duke Ellington.

Library of Congress Cataloging-in-Publication Data

Bradfield, Scott.
Greetings from earth : new and collected stories / by Scott Bradfield.
p. cm.
ISBN 0-312-14088-6 (pbk.)
1. Manners and customs—Fiction. I. Title.
PS3552.R214G7 1996
813'.54—dc20 95-46778
CIP

First published in Great Britain by Little, Brown and Company

First Picador USA Edition: April 1996
10 9 8 7 6 5 4 3 2 1

For Anne McDermid
and Brian Moore

CONTENTS

The Dream of the Wolf 1
The Darling 25
Sweet Ladies, Good Night, Good Night 43
The Wind Box 53
Closer to You 71
The Last Man That Time 80
Didn't She Know 90
In the Time of the Great Dying 102
Ghost Guessed 108
Dazzle 120
White Lamp 133
Hey Hey Hey 145
The Other Man 175
The Promise 184
The *Flash!* Kid 197
The Monster 209
Greetings From Earth 220
Unmistakably the Finest 232
Diary of a Forgotten Transcendentalist 252
The Secret Life of Houses 264
The Parakeet and the Cat 284

THE DREAM OF THE WOLF

Without the dream one would have found no occasion for a division of the world.

NIETZSCHE

'Last night I dreamed I was *Canis lupus tundrarum*, the Alaskan tundra wolf,' Larry Chambers said, confronted by hot Cream O' Wheat, one jelly donut, black coffee with sugar. 'I was surrounded by a vast white plain and sparse gray patches of vegetation. I loped along at a brisk pace, quickening the hot pulse of my blood. I felt extraordinarily swift, hungry, powerful . . .' Larry gripped his donut; red jelly squirted across his knuckles. 'My jaws were enormous, my paws heavy and callused.' He took a bite, chewed with his mouth open. 'My pelt was thick and white and warm. The cold breeze carried aromas of fox, rabbit, caribou, rodent, fowl, mollusc . . .'

'Caroline!' Sherryl Chambers reached for the damp dishcloth. 'Eat over the table, *please*. Just look at this. You've dripped cereal all over your new shoes.'

Caroline gazed up intently at her father, her chin propped against the table edge. Her fist gripped a grainy spoon.

'I heard a noise behind me and I turned.' Larry warmed his palms against the white coffee cup. 'The mouse hesitated – just for a moment – and then quickly I pounced, pinned him beneath my paw. His eyes were wide with panic, his tiny heart fluttered wildly. His fear blossomed in the air like pollen—'

'What did you do, Daddy? What did you do to the mouse?'

Larry observed the clock radio. *KRQQ helicopter watch for a Monday, March twenty-third*, the radio said. *An overturned tanker truck has traffic backed up all the way to the Civic Center . . .*

'I ate him,' Larry said. The time was eight-fifteen.

'Caroline. Finish your cereal before it gets cold.'

'But Daddy's a wolf again, Mommy. He caught a mouse and he *ate* it.'

'I'm practically certain it was the *tundrarum*,' Larry said, and pulled on his sport coat.

'Please, Caroline. I won't ask you again.'

'But I want the rest of Daddy's donut.'

'Finish your cereal. *Then* we'll discuss Daddy's donut.'

'I think I'll stop by the library again tonight.' Larry got up from the table. His spoon remained gripped by the thickening cereal like a fossil in La Brea.

'Sure, honey. And pick up some milk on the way home, will you? *Try* and remember.'

'I will,' Larry said, 'I'll try,' recalling the brilliant white ice, the warm easy taste of the blood.

'And here – bend over.' Sherryl moistened the tip of a napkin with her lips. 'There's jelly all over your face.'

'It's the blood, Daddy. It's the mouse's blood.'

'Thanks,' Larry said, and went into the living room.

Caroline watched the kitchen door swing shut. After a few moments she heard the front door open and close.

'Daddy forgot to kiss me goodbye,' she said.

Sherryl spilled pots and pans into the sink. 'Daddy's a little preoccupied this morning, dear.'

Caroline thought for a moment. The bitten jelly donut sat in the middle of the table like a promise.

'Daddy ate a mouse,' she said finally, and made a proud little flourish in the air with her spoon.

Canis lupus youngi, canis lupus crassodon, canis niger rufus, Larry thought, and boarded the RTD at Beverly and Fairfax. The wolf, he thought. The wolf of the dream, the wolf of the world. He showed the driver his pass. Wolves in Utah, Northern Mexico, Baffin Island, even Hollywood. Wolves secretly everywhere, Larry thought, and moved down the crowded aisle. Elderly women jostled fitfully in their seats like birds on a wire.

'Larry! Hey – Spaceman!'

Andrew Prytowsky waved his *Wall Street Journal*. 'Sit here.' He removed his briefcase from the window seat and placed it in his lap. 'Rest that frazzled brain of yours. You may need it later.'

'Thanks,' Larry said, squeezed into the vacant seat and recalled an exotic afternoon nap. *Canis lupus chanco*, Tibetan spring, crepuscular hour. His pack downed a goat. Blood spattered the gray dust like droplets of quivering mercury.

'*That's* earnings, Larry. *That's* reliable income. *That's* retirement security, a summer cottage, a sporty new car.' Andrew shook the American Exchange Index at him, as if reproving an unhousebroken puppy. 'Fifteen points in two weeks, just like I promised. Did you hear me? *Fifteen* points. Consolidated Plastics Ink. Plastic bullets, the weapon of the future. Cheap, easy to manufacture, minimal production overhead. You could have cut yourself a piece of that, Larry. I certainly gave you every opportunity. But then *my* word's not good enough for you, is it? You've already got your savings account, your fixed interest, your automatic teller, your free promotional albums. You've got yourself a coffin – *that's* what you've got. Fixed interest is going to bury you. Listen to me, pal. I can help. Let's talk tax-free municipal bonds for just one second—'

Larry sighed and gazed out the smudged window. Outside the Natural History Museum sidewalk vendors sold hot dogs, lemonade and pretzels while behind them ancient bones surfaced occasionally from the bubbling tar pit.

'. . . in the long run we're not just talking safety. We're talking variable income *and* easy liquidity.' Prytowsky slapped Larry's chest with the rolled-up newspaper. 'Get *with* it, Spaceman. What are you now? Late thirties, early forties? You want to spend the rest of your life with your head in the clouds? Or do you want to come back down to earth and enjoy a little of the *good* life? Your little girl – Carol, Karen, whatever. She may be four or five now, but college is *tomorrow*. *Tomorrow*, Spaceman. And you want your little girl to go to college, don't you? Well, *don't* you? Of *course* you do! Of *course*!'

The traffic light turned green, the RTD's clutch connected with a sudden sledgehammer sound. Oily gray smoke swirled outside the window.

'And what about that devilish little wife of yours? Take it from me, Spaceman. A woman's eye is *always* looking out for those greener pastures. It's not their *fault*, Spaceman it's just their *nature* . . . Hey, *Larry*.' The rolled-up newspaper jabbed Larry's side. 'You even listening to me or what?'

'Sure,' Larry said, and the bus entered Beverly Hills. Exorbitant hood ornaments flashed in the sun like grails. 'Easy liquidity, interest variations. I'll think about it. I really will. It's just I have a lot on my mind right now, that's all. I mean, I'll get back to you on all this. I really will.' *Canis lupus arabs, pallipes, baileyi, nubilis, monstrabilis*, he thought. The wolves of the dream, the wolves of the world.

'Still having those nutty dreams of yours, Spaceman? Your wife told my wife. You dream you're a dog or something?'

'A wolf. *Canis lupus*. It's not even the same subspecies as a dog.'

'Oh.' Andrew discarded his newspaper under his seat. 'Sure.'

'Wolves are far more intelligent than any dog. They're fiercer hunters, loyaler mates. Their social organization alone—'

'Yeah – right, Spaceman. I stand corrected. I'll bet in your dreams you really raise hell with those stupid dogs – hey, Larry, old pal?' Andrew said, and disboarded with his briefcase at Westwood Boulevard.

As the bus approached 27th Avenue Larry moved back through the crowd of passengers who stood and sat about with newspapers, magazines and detached expressions as they vacantly chewed Certs, peanuts from a bag, impassive bubble gum, like a herd of grazing buffalo while the wolf, the wolf of Larry's mind, roamed casually among them, searching out the weak, the sickly, the injured, the ones who always betrayed themselves with brief and anxious glances – the elderly woman with the aluminum walker, the gawky adolescent with the bad complexion and crooked teeth. Wolves in Tibet, Montana, South America, Micronesia, Larry thought, disembarked at 25th Avenue and entered Tower Tire and Rubber Company. He showed his pass to the security guard, then rode the humming elevator to the twelfth floor. When Larry stepped into the foyer the secretaries, gathered around the receptionist's desk, exchanged quick significant glances like secret memoranda. Larry heard them giggling as he disappeared into the maze of high white partitions that organized office cubicles like discrete cells in an ant farm.

Larry entered his office.

'Ready for Monday?' Marty Cabrillo asked.

Larry hung his coat on the rack, turned.

The marketing supervisor stood in front of Larry's aluminum bookshelf, gazing aimlessly at the spines of large gray Acco-Grip binders. 'Frankly,' Marty said, 'I'd rather be in Shasta. How was your weekend?'

'Fine, just fine,' Larry said, sat down at his desk and opened the top desk drawer.

'I thought I'd drop by and see if the Orange County sales figures were in yet. Didn't mean to barge in, you know.'

'Certainly. Help yourself.' Larry gestured equivocally with his right hand, rummaged in the desk drawer with his left.

'Ed Conklin called from Costa Mesa and said he still hasn't received the Goodyear flyers. I told him no problem – you'd get right back to him. All right?'

'Right.' Larry slammed shut one drawer and pulled open another. 'No problem. Here we are . . .' He removed a large faded-green hardcover book. One of the book's corners was bloated with dog-eared pages. Larry wiped off dust and bits of paper against his trousers. *The Wolves of North America: Part 1, Classification of Wolves*.

Marty propped one hand casually in his pocket. 'I hope you don't take this the wrong way or anything, Larry . . . I mean, I'm not trying to pull rank on you or anything. But maybe you could try being just a little bit more careful around here the next few weeks or so. Think of it as a friendly warning, okay?'

Larry looked up from his book.

'It's not me, Larry.' Marty placed his hand emphatically over his heart. 'You know me, right? But district managers are starting to complain. Late orders, unitemized bills, stuff like that. *Harmless* stuff, really. Nothing I couldn't cover for you. But the guys upstairs aren't so patient – that's all I'm trying to say. I'm just trying to say it's my job, too. All right?'

Finally Larry located the *tundrarum*'s subspecies guide. *Type locality: Point Barrow, Alaska. Type Specimen: No. 16748, probably female, skull only, US National Museum; collected by Lt P. H. Ray . . .*

'But for God's sake don't take any of this personal or anything. It's not really serious. Everybody has their off days – it's just the way things go. People get, well, *distracted*.'

'I knew it.' Larry pointed at the page. 'Just what I thought. Look –

tundrarum is "closely allied to *pambisileus*." Exactly as I suspected. The dentition was a dead giveaway.'

Marty fumbled for a cigarette from his shirt pocket, a Bic lighter from his slacks. 'Well,' he said, and took a long drag from his Kool. Then, after a moment, 'You know, Larry, Beatrice and I have always been interested in this ecology stuff ourselves. You should visit our cabin in Shasta sometime. There's nothing like it – clean air, trees, privacy. We even joined the Sierra Club last year . . . But look, I could talk about this stuff all day, but we've *both* got to get back to work, right?' Marty paused outside the cubicle. 'We'll get together and talk about it over lunch sometime, okay? And maybe you could drop the sales figures by my office later? Before noon, maybe?'

That night Larry returned home after the dinner dishes had been washed. He glanced into Caroline's room. She was asleep. Stuffed wolves, cubs and one incongruous unicorn lay toppled around her on the bed like dominoes. He found Sherryl in the master bedroom, applying Insta-Curls to her hair and balancing a black rectangular apparatus in her lap.

Larry sat on the edge of the bed, glimpsed himself in the vanity mirror. He had forgotten to shave that morning. His eyes were dark, sunken, feral. (The lone wolf lopes across an empty plain. Late afternoon, clear blue sky. The pale cresent moon appears on the horizon like a specter. Other wolves howl in the distance.)

Larry turned to his wife. 'I went all the way out to the UCLA Research Library, then found out the school's between quarters. The library closed at five.'

'That's too bad, dear. Would you plug that in for me?'

Sherryl pulled a plastic cap over her head. Two coiled black wires attached the cap to the black rectangular box. Larry connected the plug to the wall socket and the black box began to hum. Gradually the plastic cap inflated. 'Larry, I wish I knew how to phrase this a bit more delicately, but it's been on my mind a lot lately.' Sherryl turned the page of a K Mart Sweepstakes Sale brochure. 'You may not believe this, Larry, but there are actually people in this world who like to talk about some things besides *wolves* every once in a blue moon.'

Larry turned again to his reflection. He had forgotten to finish Cabrillo's sales figures. Tomorrow, he assured himself. First thing.

'I remember when we had decent conversations. We went out

occasionally. We went to movies, or even dancing. Do you remember the last time we went out together – I mean, just out of the *house*? It was that horrid PTA meeting last fall, with that dreadful woman – the hunchback with the butterfly glasses, you remember? Something about a rummage sale and new tether poles? Do you *know* how long ago that was? And frankly, Larry, I wouldn't call that much of a night *out*.'

Larry ran his hand lightly along the smooth edge of the humming black box. 'Look, honey. I know I get a little out of hand sometimes . . . I *know* that. Especially lately.' He placed his hand on his forehead. A soft pressure seemed to be increasing inside his skull, like an inflating plastic cap. 'I've been forgetful . . . and I realize I must seem a little nutty at times . . .' The wolves, he thought, trying to strengthen himself. The call of the pack, the track of the moon, the hot quick pulse of the blood. But the wolves abruptly seemed very far way. 'I know you don't understand. *I* don't really understand . . . But these aren't just dreams. When I'm a wolf, I'm *real*. The places I see, the feelings I feel – they're *real*. As real as I am now, talking to you. As real as this bed.' He grasped the king-size silk comforter. 'I'm not making all this up . . . And I'll *try* to be a little more thoughtful. We'll go out to dinner this weekend, I promise. But try putting up with me a little longer. Give me a little credit, that's all . . .'

Sherryl glanced up. She took the humming black box from his hand.

'Did you say something, hon?' She patted the plastic cap. 'Hold on and I'll be finished in a minute.' She turned another page of the brochure. Then with a heavy red felt marker, she circled the sale price of Handi-Wipes.

Larry walked into the bathroom and brushed his gleaming white teeth.

'Last night I dreamed of the Pleistocene.'

'Where is that, Daddy?'

'It's not a place, honey. It's a time. A long time ago.'

'You mean dinosaurs, Daddy? Did you dream you were a *dinosaur*?'

'No, darling. The dinosaurs were all gone by then. I was *canis dirus*, I think. I'll check on it. The tundra was far colder and more desolate than before. The sky was filled with this weird, reddish glow I've never seen before, like the atmosphere of some alien planet. Ice was everywhere. Three of us remained in the pack. My mate had died the previous night beneath a shelf of ice while the rest of us huddled around to keep her warm. Dominant, I led the others across the white ice, my tail slightly erect. We were terribly cold, tired, hungry . . .'

'Weren't there any mice, Daddy? Or any snails?'

'No. We had traveled for days. We had discovered no spoor. Except one.'

'Was it a deer, Daddy? Did you kill the deer and eat it?'

'No. It was Man's spoor. We were seeking an encampment of men.' He turned. Sherryl was beating eggs into a bowl and watching David Hartman on the portable television. 'Sherryl, that was the strangest part. I've read about it, anthropologists have suggested it – a prehistoric, communal bond between man and wolf. We weren't afraid. We sought shelter with them, food, companionship, allies in the hunt.'

Larry watched his wife. After a moment she said, 'That's nice, dear.'

David Hartman said, 'Later in this half hour we'll be meeting Lorna Backus to discuss her new hit album, and then take an idyllic trip up the coast to scenic New Hampshire, the Granite State, as part of our "States of the Union" series. Please stay with us.'

'I've always wanted to live in New Hampshire,' Sherryl said.

Every day on his way home from work Larry stopped at the Fairfax branch library. Many of the books he needed he had to request through interlibrary loan. He read Lopez's *Of Wolves and Men*, Fox's *The Soul of the Wolf*, Mech's *The Wolf: The Ecology and Behaviour of an Endangered Species*, Pimlott's *The World of the Wolf*, Mowat's *Never Cry Wolf*, Ewer's *The Carnivores*, and the pertinent articles and symposiums published in *American Zoologist*, *American Scientist*, *Journal of Zoology*, *Journal of Mammalogy* and *The Canadian Field Naturalist*. Sherryl pulled the blankets off the bed one day and three books came loose, thudding on to the floor. 'I'd really appreciate it, Larry, if you could start picking up after yourself. It's bad enough with Caroline. And just look – this one's almost a month overdue.' Larry returned them to the library that night, checked out three more, and xeroxed the 'Canids' essay in *Grzimek's Animal Life Encyclopedia*.

On the way out the door he noticed a three-by-five file card tacked to the Community Billboard. *Spiritual Counseling, Dream Analysis, Budget Rates, Free Parking*. Her name was Anita Louise. She lived on the top floor of a faded Sunset Boulevard brownstone, and claimed to be circuitously related to Tina Louise, the former star of *Gilligan's Island*. Her living room was furnished with tattered green lawn chairs and orange-crate bookshelves. She required a personal item; Larry handed her his watch. She closed her eyes. 'I can see the wolf now,' she said. Her fingers smudged the watch's crystal face, wound the stem, tested the flexible metal band. 'While he leads you through the forest of life, he warns you of the thorny paths. When the time comes, he will lead you into Paradise.'

'The wolf doesn't guide me,' Larry said. 'I *am* the wolf. Sometimes *I* am the guide, the leader of my pack.'

'The ways of the spirit world are often baffling to those unlearned in its ways,' Anita told him. 'I take Visa and MasterCard. I take personal checks, but I need to see at least two pieces of ID.'

Before he left, Larry reminded her about his watch.

'I don't know, Evelyn. I really just don't know. I mean, I *love* Larry and all, but you can't imagine how difficult life's been around here lately – especially the last few months.' Sherryl held the telephone receiver with her left hand, a cold coffee cup with her right. She listened for a moment. 'No, Evelyn, I don't think *you* understand. This isn't a hobby. It's not as if Larry was collecting stamps, or a *bowler* or something. I could understand that. *That* would be understandable. But all Larry talks about anymore is wolves. Wolves this and wolves that. Wolves at the dinner table, wolves in bed, wolves even when we're driving to the market. Wolves are everywhere, he keeps saying. And honestly, Evelyn, sometimes I almost believe him. I start looking over my shoulder. I hear a dog bark and I make sure the door's bolted . . . Well, of *course* I try to be understanding. I'm trying to tell you that. But I have to worry about Caroline too, you know . . . Well, listen for a minute and I'll tell you what happened yesterday. We're sitting at breakfast, you see, and Larry starts telling Caroline – a four-year-old girl, remember – how he's off in the woods somewhere, God only knows *where*, and he meets this female dog and, well, I can't go on . . . No, I simply can't. It's too embarrassing . . . No, Evelyn. You've completely missed the point. It's mating season, get it? And Larry starts going into explicit detail . . . Well, maybe. But that's not even the worst part . . . Hold *on* for one second and I'll tell you. They, well, I don't know how to phrase this delicately. They get *stuck* . . . *No*, Evelyn. Honestly, sometimes I don't think you're even listening to me.

They get stuck *together*. Can you believe that? What am I supposed to say? Caroline's not going to outgrow a trauma like this, though. I can promise you that.' Sherryl heard the kitchen door opening behind her. 'Hold on, Evelyn,' she said, and turned.

Caroline blocked the door open with her foot. 'What are you talking about?' Her hand gripped the plastic Pez dispenser. Wylie Coyote's head was propped back by her thumb, and a small pink lozenge extruded from his throat.

'It's Evelyn, dear. We're just talking.'

Caroline's lips were flushed and purple; purple stains speckled her white dress. She thought for a moment, took the candy with her teeth and chewed. Finally she said, 'I think somebody may have spilled grape juice on one of Daddy's wolf books.'

Larry read Guy Endore's *The Werewolf of Paris*, Hesse's *Steppenwolf*, Rowland's *Animals With Human Faces*, Pollard's *Wolves and Werewolves*, Lane's *The Wild Boy of Aveyron*, Malson's *Wolf Children and the Problem of Human Nature*. Marty gave him the card of a Jungian in Topanga Canyon who sat Larry in a plush chair, said 'archetype' a few times, informed him that *everyone* is fascinated with evil, sadism, pain ('It's perfectly normal, perfectly *human*'), recommended Robert Eisler's *Man into Wolf*, charged seventy-five dollars and offered him a Valium prescription with refill. 'But when I'm a wolf, I never know evil,' Larry said as he was ushered out the door by a blond receptionist. 'When I'm a wolf, I know only peace.'

'I don't know, Larry. It just gives me the creeps,' Sherryl said that night after Caroline was in bed. 'It's *weird*, that's what it is. Bullying defenseless little mice and deer that never hurt anybody. Talking about killing, and blood, and ice – and particularly at *breakfast*.'

Larry was awake until two a.m. watching *The Wolf Man* on Channel Five. Claude Raines said, 'There's good and evil in every man's soul. In this case, the evil takes the shape of a wolf.' No, Larry thought, and read Freud's *The Case of the Wolf-Man*, the first chapter of Mack's *Nightmares and Human Conflict*. No. Then he went to bed and dreamed of the wolves.

*

The wolf spirit has always been considered very *wakan*,' Hungry Bear said, his feet propped on his desk. He poked out his cigarette against the rim of the metal wastebasket, then prepared to light another. 'Most tribes believe the wolf's howl portends bad things. The Lakota say, "The man who dreams of the wolf is not really on his guard, but the man haughtily closes his eyes, for he is very much on his guard." I don't know what that means, exactly, but I read it somewhere.' Hungry Bear refilled his Dixie Cup with vin rosé. His grimy T-shirt was taut against his large stomach; a band of pale skin bordered his belt. He wore a plaid Irish derby atop his braided hair. 'I try to do a good deal of reading,' he said, and fumbled in his diminished pack of Salems.

'So do I,' Larry said. 'Maybe you could recommend—'

'I don't think the wolf was ever recognized as any sort of deity, but I could be wrong.' Hungry Bear was watching the smoke unravel from his cigarette. 'But still, you shouldn't be too worried. It's very common for animal spirits to possess a man. They use his body when he's asleep. When he wakes, he can't remember anything . . . oh, but wait. That's not quite right, is it? You said you *remember* your dreams? Well, again, I could be wrong. I guess you *could* remember. Sure, I don't see why not,' Hungry Bear said, and poured more vine rosé.

'*I* inhabit the body of the *wolf*,' Larry said, beginning to lose interest, and glanced around the cluttered office. The venetian blinds were cracked and dusty, the floors littered with tattered men's magazines, empty wine bottles and crumpled cigarette packs. After a moment he added, 'I don't even know what I should call you. *Mister* Bear?'

'No, of course not.' Hungry Bear waved away the notion, dispersing smoke. 'Call me Jim. That's my real name. Jim Prideux. I took Hungry Bear for business purposes. If you remember, Hungry Bear was the brand name of a terrific canned chili. It was discontinued after the war, though, I'm afraid.' He checked his shirt pocket. 'Do you see a pack of cigarettes over there? Seems I'm running short.'

'You're not Indian?' Larry asked.

'Sure. Of course I'm Indian. One-eighth pure Shoshone. My great-grandmother was a Shoshone princess. Well, maybe not a princess, exactly. But *her* father was an authentic medicine man. I've inherited the gift.' Jim Prideux rummaged through the papers on his desk. 'Are you sure you don't see them? I'm sure I bought a pack less than an hour ago.'

*

'This is very nice,' Sherryl said, and swallowed her last bite of red snapper. She touched her lips delicately with the napkin. 'It's *so* nice to get out of the house for a change. You wouldn't know how much.'

'Sure I would, darling,' Andrew Prytowsky said, and poured more Chenin Blanc.

'No, I don't think you would, Andy. Your wife, Danielle, is *normal.* You wouldn't know what it's like living with someone as . . . well, as *unstable* as Larry's been acting lately.'

'I'm sure it's been very difficult for you.'

'Marty Cabrillo, Larry's boss at work, got Larry in touch with a doctor, a *good* doctor. Larry visits him *once* and then tells me he isn't going anymore. I say to Larry, don't you think he can help you? And Larry says no, he can't, he can't help him at all. He says the doctor is *stupid.* Can you believe that? I say to Larry, this man has a *Ph.D.* I don't think you can just call a man with a Ph.D. *stupid.* And so then Larry says *I* don't know what *I'm* talking about, either. Larry thinks he knows more than a man with a Ph.D. That's what Larry thinks.'

'Here. Why don't you finish it?' Andrew put down the empty bottle and flagged the waiter with his upraised MasterCard.

'I'm sorry, Andy.' Sherryl dabbed her eyes with the napkin. 'It's just I'm so shook up lately. All I ever asked for was a normal life. That's not too much, is it? A nice home, a normal husband. Someone who could give me a little help and support. Is that too much to ask? Is it?'

'Of course not.' Andrew signed the check. After the waiter left he said, 'I'm glad we could do this.'

Sherryl folded her napkin and replaced it on the table. 'I'm glad you called. This was very nice.'

'We'll do it again.'

'Yes,' Sherryl said. 'We should.'

Two weeks later Larry returned home from work and found the letter on the kitchen table.

Dear Larry,
I know you're going to take this the wrong way and I only hope you realize Caroline and I still care about you but I've thought about this a lot and even sought professional counseling on one

occasion and I think it's the only solution right now at this moment in our lives. Especially for Caroline who is at a very tender age. Please don't try calling because I told my mother not to tell you where we are for a while. Please realize I don't want to hurt you and this will probably be better for both of us in the long run, and I hope you make it through your difficulties and I'll think good thoughts for you often.

Sherryl

'You can't just keep moping around, Larry. Things'll get better, just you wait. I sense big improvements coming in your life. But first you've *got* to start being more careful around the office.' Marty sat on the edge of Larry's desk. He pulled a string of magnetized paper clips in and out of a clear plastic dispenser. 'Did I tell you Henderson asked about you yesterday? Asked about you *by name*. Now, I'm not trying to make you paranoid or anything, but if Henderson asked about you, then you can bet your socks the *rest* of the guys in Management have been tossing your name around. And Henderson's not a bad guy, Larry. I'm not suggesting that. But there's been a sincere ... a sincere *concern* about your performance around here lately. And don't think I don't understand. Really, Larry, I'm very sensitive to your position. Beatrice and I came close to breaking up a couple of times ourselves – and I don't know *what* I'd do without Betty and the kids. But you've got to keep your chin up, Buddy. Plow straight ahead. And remember – I'm on *your* side.'

At his desk, Larry made careful, persistent marks on a sheet of graph paper. The frequency of dreams had increased over the past few weeks: the line on the graph swooped upward. Often three, even four times a night he started awake in bed, clicked on the reading lamp and reached for a pen and notepad from the end table, quickly jotting down terrain and subspecies characteristics while the aromas of forest, desert and tundra were displaced by the close, stale odors of grimy bedsheets, leftover Swanson frozen dinner entrées, and Johnson's Chlorophyll-Scented Home Deodorizer.

'I'm really sincere about this, Larry. I can't keep covering for you. I need some assurances, I need to start seeing some real *effort* on your part. You're going to start seeing Dave Boudreau on the third floor. He's our employee stress counselor – but that doesn't mean he's like a shrink or anything, Larry. I know how you feel about *them*. Dave Boudreau's just a regular guy like you and me who happens to have a lot of experience with these sorts of problems. You and Sherryl, I mean. All right, Larry? Does that sound fair to you?'

'Sure, Marty,' Larry said, 'I appreciate your help, I really do,' and peeled another sheet from the Thrifty pad. Abscissa, he thought: real time. Ordinate: dream time. At the top of the page he scribbled *Pleistocene*.

'I'm dreaming now more than ever,' Larry told Dave Boudreau the following Thursday. 'Sometimes half a dozen times each night. Look, I've kept a record—' Larry opened a large red loose-leaf binder, flipped through a sheaf of papers and unclamped a sheet of graph paper. 'There, that's last Friday. Six times.' He held the sheet of paper over the desk, pointing at it. 'And Sunday – *seven* times. And that's not even the significant part. I haven't even got to *that* part yet.'

Dave Boudreau sat behind his desk and rocked slightly in a swivel chair. He glanced politely at the statistical chart. Then his abstract gaze returned to Tahitian surf in a framed travel poster. He heard the binder clamp click again.

Larry pulled up his chair until the armrests knocked the edge of the desk. 'Increasingly I dream of the Pleistocene, the Ice Age. The Great Hunt, when man and wolf hunted together, bound by one pack, responsible to one community, seeking their common prey across the cold ice, beneath the cold sun. Is *that* something? Is that one hell of an archetype or what?'

Casually Boudreau opened the manila folder on his desk.

CHAMBERS, LAWRENCE
SUPPLIES AND SERVICES DEPARTMENT
BORN: 3-6-45 EYES: BLUE

'And don't get me wrong. I'm just kidding about that archetype stuff. That's not even close, that's not even in the same ballpark. These aren't memories, for chrissakes. When I dream of the wolf, I *am* the wolf. I've been wolves in New York, Montana and Beirut. It's as if time and space, dream and reality, have just *opened up*, joined me with everything, everything *real*. I'm living the *one life*, understand? The life of the hunter and the prey, the dream and the world, the blood and the spirit. It's really spectacular, don't you think? Have you ever heard anything like it?'

In the space reserved for Counselor's Comments Boudreau scribbled 'wolf nut,' and underlined it three times.

When Larry arrived at work the following Monday the security guard took his ID card and, after consulting his log, asked him to please wait one moment. The guard picked up his phone and asked the operator for Personnel Management. 'This is station six. Mr Lawrence Chambers has just arrived.' The guard listened quietly to the voice at the other end. He snapped his pencil against the desk in four-four time.

Finally he put down the phone and said, 'I'm sorry. I'll have to keep your card. Would you please follow me?'

They walked down the hall to Payroll. Larry was given his final paycheck and, in a separate envelope, another check for employee minimum compensation.

By the time Larry returned home it was still only ten a.m. He cleared the old newspapers from the stoop, unbound and opened the whitest, most recent one. He read for a few minutes, then refolded the paper and placed it with the others beside the fireplace. He picked up Harrington and Paquet's *Wolves of the World* and put it down again. He got up and walked to the kitchen. Dishes piled high in the sink, four full bags of trash. The few remaining dishes in the dishwasher were swirled with white mineral deposits. In the refrigerator he found a garlic bulb with long green shoots, an empty bottle of Worcestershire sauce and an egg. He drank stale apple juice from the plastic green pitcher, then continued making his rounds. In the bathroom: toothpaste, toothbrush, comb, water glass, eye drops, Mercurochrome, a stray bandage, Sherryl's pH-balanced Spring Mountain shampoo, his electric razor. All the clothes and toys were gone from Caroline's room. Over the bed the poster of a wolf gazed down at him, its eyes sharp, canny, primitively alert.

He tried to watch television. People won sailboats and trash compactors on game shows, cheated one another and plotted financial coups on soap operas. After a while he got up again and returned to the bathroom, opened the medicine cabinet. Johnson's Baby Aspirin, an old stiffened toothbrush, mouthwash, a bobby pin. High on the top shelf he found Sherryl's Seconal in a childproof bottle. He took two. Then he got into bed.

Sometime after dawn he dreamed again of the wolves, but this time the dream was fragmentary and detached. He viewed the wolves from very far away. From atop a high bluff, perhaps, or hidden behind some bushes like Jane Goodall. The wolves moved down into the gully and paused before a small stream, drinking. Two cubs splashed and chased one another through the puddles. Other wolves observed them dispassionately. The sun was going down. Larry woke up. It was just past six a.m.

He stayed indoors throughout the day. In the evening he might walk to the corner Liquor Mart to cash a check and purchase milk, scotch, Stouffer's frozen dinners. Sometimes, remembering Sherryl and Caroline, he turned the television up louder. It wasn't their physical presence he missed (he could hardly recall their faces anymore) but rather their noise: the clatter of dishes, the inconstant whir and jingle of mechanical toys. Soundless, the air seemed thinner, staler, more oppressive, as if he were sealed inside an airtight crystal vault. The silence invested everything – the walls, the furniture, the diminishing vial of Seconal, the large empty bedrooms, even the mindless chatter of the Flintstones on television. He drank his beer beside the front window and watched the dust swirl soundlessly in the soundless shafts of light, recalling the wolves and the soundless expanse of white ice where not only the noise but even the aromas and textures of the landscape seemed to be leaking from the dreamlike atmosphere, as if from the cracks in some domed underwater city. In the mornings, now, he hardly recalled his dreams at all anymore. Sporadic glimpses of wolf, prey, sky, moon, interspliced meaninglessly like the frames of some surrealist montage. He smoked three packs of cigarettes a day, just to give his hands something to do. The scotch and Seconal compelled him to take so many naps during the day that he couldn't sleep at night. Wolves, he thought. Wolves in Utah, Baffin Island, even Hollywood. Wolves secretly everywhere . . . Eventually the dreams disappeared entirely. Sleep became a dark visionless place where nothing ever happened.

The Seconal, he thought one morning, and departed for the library. He squinted at the sunlight, staggered occasionally. People looked at him. A book entitled *Sleep* by Gay Gaer Luce and Julius Siegal confirmed his suspicions. Alcohol and barbiturates suppressed the dream stage of sleep. He returned home and poured the scotch down the sink, the remaining Seconal down the toilet. He lay in bed throughout the afternoon, night and following morning. He tossed and turned. He couldn't keep his eyes closed more than a minute. His heart palpitated disconcertingly. He tried to remember the wolf's image, and remembered only pictures in books. He tried to recall the prey's hot, steaming blood, and tasted only yesterday's Chicken McNuggets. He wanted the map of the sky, and found only the close, humid rectangle of the bedroom. He got up and went into the living room. It was night again. In order to dream, he must sleep. In order to regain the real, he must dispel the illusion: newspapers, furniture, unswept carpets, Sherryl's letter, Caroline's toys, easy liquidity, magazines and books. He realized then that evil was not the wolf, but rather the wolf's disavowal. Violence wasn't something in nature, but rather something in nature's systematic repression. Madness isn't the dream, but rather the world deprived of the dream, he thought, selected a stale pretzel from the bowl, chewed and gazed out the window at the dim, empty streets below, where occasional streetlamps illuminated silent, unoccupied cars parked along the curbs. The moon made a faint impression against the high screen of fog. A distant siren wailed, a dog barked and in their homes the population slept fitfully, often aided by Seconal and Valium, descending through soft penetrable stages of sleep, seeking that fugitive half-world in which they struggled to dream beneath the repressive shadows of the real.

A few weeks after signing Larry Chamber's termination notice, Marty Cabrillo took his wife to Shasta. 'Two weeks alone,' he promised her. 'We'll leave the kids with your mother. Just the two of us, the trees, candlelight dinners again, just like I always said it would be.' But Marty said nothing during the long drive. Beatrice put her arm around him and he shrugged at her. 'Please,' he said. 'I can't get comfortable.' At the cabin they sat out on the sun deck. Marty held paperbacks and turned pages. Beatrice read *People* magazine. After only a few days they returned home. 'I'm sorry, honey,' Marty said to her. 'I'll make it up to you. I promise.'

'What's the matter with you lately?'

'Nothing. Just things on my mind.'

'Work?'

'Sort of.'

After a while Beatrice said, 'Larry,' folded her arms and gazed out the window at Ventura car lots.

The following Sunday Marty drove to Ralph's in Fairfax, loaded four bags of groceries into his Toyota station wagon and drove to

Larry's house on Clifton Boulevard. The front yard was brown and overgrown. Aluminum garbage cans, streaked with rust, lay overturned in the alley. Dormant snails studded the front of the house, their slick intricate trails glistening in the sunlight. Marty knocked, rang the bell a few times. The door was ajar and he pushed it open. A pyramid of bundled newspapers blocked the door, permitting him just to squeeze through. In the living room, torn magazines and moldy dishes lay strewn across the sofa, chairs and floor. The telephone receiver was off the hook, wailing faintly like a distant, premonitory siren. At first the room seemed oddly disproportionate, as if the furniture had all been rearranged. Then he noticed Larry asleep on the middle of the floor, his head propped by a sofa cushion, his arm wrapped around a leg of the coffee table. 'He must've lost eighty, ninety pounds,' Marty told Beatrice later that night. 'His clothes stank, he hadn't shaved or washed in I don't know how long. And all I could think looking at him there was it's all *my* fault. *I* was responsible. Me, Marty Cabrillo.'

Marty followed the ambulance to St John's, wishing they would run the siren. 'Dehydration,' the doctor told him, while Marty paid the deposit on a private room. Larry lay in a stiff, geometric white bed, a glucose bottle hanging beside him, a white tube connected to his arm by white adhesive tape. Every so often the glucose bubbled. 'We'll bring him along slow, have him eating solid food in a couple days. I think he'll be all right,' the doctor said, and handed Marty another form to sign.

'It's all my fault,' Marty said when Larry regained consciousness the following morning. 'Look, I brought you some books to read. And the flowers – they're from Sherryl. Beatrice got in touch with her last night and she's on her way here right now. The worst is over, pal. The worst is all behind you.'

Later Sherryl told him, 'We missed you. Caroline missed you. *I* missed you. Oh, Larry. You just look so *aw*ful.' Sherryl laid her head in Larry's lap and cried, hugging him. Silently Larry stroked her long blond hair. Sherryl had been staying with her sister in Burbank, working as a secretary at one of the studios. Her boss was a flushed, obese little man who put his hand on her knee while she took dictation, or snuck up behind her every once in a while and gave her a sharp pinch. 'Loosen up, relax. Life's short,' he told her. Caroline hated her new nursery school and cried nearly every day. Sherryl's sister had begun bringing classified pages home, pointing out to her the best bets on her own apartment. Andy had promised to help out, but every time she called his office his secretary said he was still out of town on business. And then one of the Volvo's tires went flat, and in all the rush of moving she realized she had misplaced her triple-A card, and so she just started crying, right there on the side of the freeway, because it seemed as if nothing, nothing ever went right for her anymore.

'We need you, Larry,' Sherryl said. 'You need us. I'm sorry what happened, but I always loved you. It wasn't because I didn't love you. And Marty thinks he can get your old job back—'

Marty leaned forward, whispered something.

'He says he's certain. He's certain he can get it back. Did you hear, honey?' Everything's going to be all right. We're all going to be happy again, just like before.'

Sherryl brought Caroline home a month later.

'Is Daddy home?' Caroline asked.

'He's at work now, honey. But he'll be back soon. He's missed you.'

Caroline waited to be unbuckled, climbed out of the car. The front yard was green and delicate, the house repainted yellow. The place seemed only dimly familiar, like the photograph Mommy showed her of where she lived when she was born.

'All your toys are in your room, sweetheart. Be good and play for a while. Mommy'll fix dinner.'

Caroline's room had been repainted too. Over her bed hung a bright new Yosemite Sam poster. She opened the oak toy chest. The toys were boxed and neatly arranged, just like on shelves at the store. She went into the bedroom and looked at Daddy's bookcase. The large picture books were gone, along with their photographs of wolves and deer and rabbits and forests and men with rifles and hairy, misshapen primitive men. Bent paperbacks had replaced them. The covers depicted beautiful men and women, Nazi insignia, secret dossiers, demonic children, cowboys on horses, murder weapons.

She heard the front door open. 'Hi, honey. Sorry I'm late. I ran into Andy Prytowsky on the bus – remember him? I introduced you at a party last year. Anyway, I told him I'd drop by his office tomorrow. I figure it's time we started some sort of college fund for Caroline. I'm pretty excited about it. Andy says he can work us a nice little tax break, too. Oh, and look what else. I bought us some wine. For later.'

Caroline walked halfway down the hall. Mommy and Daddy stood at the door, kissing.

'There she is. There's my little girl.'

Daddy picked her up high in the air. His face seemed strange and unfamiliar, like the front of the house.

'So how have you been, sweetheart?' Daddy put her down.

'I'll finish dinner,' Sherryl said.

'Come and sit down.' Daddy led her to the sofa. 'Tell me what you've been up to. Did you have fun at Aunt Judy's?'

Caroline picked at a scab on her knee. 'I guess.'

'What do you want to do? I thought we'd go to a movie later. Would you like that?'

Caroline clasped her hands in her lap. Here is the church, and here is the steeple. When you open the doors you see all the people.

'What should we do right now? Do you want to play a game? Do you want me to read you one of your Dr Seuss books?'

Caroline thought for a while. Daddy's large, rough hand ran through her hair, snagging it. Delicately, she pushed his hand away.

'I want to watch television,' she said after a while.

Three nights each week Larry went to the YMCA with Marty. Sherryl began subscribing to *Sunset Magazine*, and over dinner they discussed a new home, or at least improvements on their present one. Finally Marty suggested they buy into his Shasta property. 'Betty and I don't make it up there more than three or four times a year. The rest of the time it'd be all yours.' Larry took out a second mortgage, paid Marty a lump sum and began sharing the monthly payments. The first few months they drove up nearly every weekend. Then Larry received a promotion which required him to make weekly trips to the Bakersfield office. 'I'm really bushed from all this driving,' he told Sherryl. 'We'll try and make Shasta *next* weekend.' Caroline started grade school in the fall. Sherryl joined an ERA support group and was gone two nights a week. Occasionally Larry spent the night in Bakersfield, and drove from there directly to work the next morning.

'All I told Conklin was I've got a merchandise deficit from his store three months in a row. It wasn't like I called him a thief or anything. I just wanted an explanation. I'm entitled to that much, don't you think? It's my job, right?'

'I'm sure he didn't mean it, Larry. He was probably just upset.' Sherryl sat on the sofa, smoking a cigarette.

'I'm sure he *was* upset. I'm sure he *was*.' Larry sat at the dining room table. The table was covered with inventories, company billing statements and large gray Acco-Grip binders. His briefcase sat open on the chair beside him. 'And now *I'm* a little upset, all right? Is that all right with you?'

'I'm sure you are, Larry. I was just saying maybe he didn't mean it, that's all. That's all I said.'

Larry put down his pencil. 'No. I don't think that's all you said.'

Sherryl looked at the *TV Guide* on the coffee table, considered picking it up. Then she thought she heard Caroline's bedroom door squeak open down the hall.

'What you said was I'm imagining things. Isn't *that* what you said?'

Sherryl crushed out her cigarette. 'Larry, I really wish you'd stop snapping at *me* every time you're mad at somebody.' She got up and went to the end of the hall. 'Caroline? Aren't you supposed to be in bed?'

Caroline's door squeaked shut. Sherryl watched the parallelogram of light on the hall floor diminish to a fine yellow line. 'And turn off those lights, young lady. You heard me. Right now,' Sherryl said. In high school Billy Mason had a crush on me, she thought, but I wouldn't give him the time of day. That morning she had seen Billy's picture on the cover of *Software World* at the supermarket.

'What I mean is, Larry, is that you're not the *only* person who's had a bad day sometimes—'

Sherryl was turning to face him when the telephone rang.

'Sometimes *my* day hasn't been that hot either,' she said, and retreated to the telephone, picked up the receiver. 'Hello?'

'Hi. Hello,' the voice said. 'I was hoping, well, I mean I didn't want to disturb anybody, but I wondered if Mr Chambers was in. Mr *Larry* Chambers, I think? Have I got that right?'

'This is his wife. Who's this?'

'Who is it?' Larry asked, picking up his pencil and jotting a number on his notepad.

Sherryl gazed expressionlessly over Larry's head at the dining-room window and, beyond, the 7-Eleven marquee. The voice on the phone filled her ears like radio static. ' – I mean, I just had the article here a moment ago, let me see . . . Look, tell him Hungry Bear called, and by the time he calls back I'll find the article – wait, in fact here it is right here – no, sorry, *that's* not it. But still, tell him Jim called. Jim Prideux—' Sherryl looked around the kitchen. She had forgotten to clean up after dinner. The sink was filled with dirty dishes, the countertop littered with bread crumbs. Stray Cheerios from that morning's breakfast had attached themselves like barnacles to the Formica table. She pulled up a chair and sat down, feeling suddenly tired. There was a television movie she had been looking forward to all week, and now, by the time she finished her cleaning, the show would practically be half over. She felt like saying to hell with it, to hell with all of it. She just wanted to go to bed. To hell with Larry, Caroline, the dishes, the vacuuming – every damn bit of it. The voice buzzed inconstantly in her ear like a mosquito, something about wolves, Navajo deities, sacred totems, irrepressible dreams of wolves, he wasn't exactly sure . . . Wolves wolves wolves, wolves everywhere, she thought, and strenghened her grip on the receiver. 'Listen to me,' she said. 'Listen to me, Mr Bear, or Mr Prideux, or Mr Whoever You Are. Listen to me for just one minute, and I'll say this as *nicely* as I can. Please don't call here anymore. Larry's not interested, *I'm* not interested. Frankly, Mr Bear, I don't think *anybody's* interested. I don't think anybody's really interested at all.'

In Sherryl's dream the men and wolves loped together across the white plain. Larry was there, and Caroline, and Andy and Evelyn and Marty and Beatrice. Sherryl recognized the mailman, the newspaper boy, supermarket employees, former boyfriends and lovers. Even her parents were there, keeping pace with wolves under the cold moonlight. Everybody was dressed as usual: The men wore slacks, ties, cuff links and starched shirts; the women skirts, blouses, jewelry and high heels. Caroline carried one of her toys, Andy his briefcase, Marty his racquetball racket and Larry one of his largest gray Acco-Grip binders. Sherryl raised a greasy spatula in her right hand, a tarnished coffeepot in her left. We forgot to schedule Caroline's dental appointment, she told Larry. When I was a child you treated me as if I was stupid, she told her father, but I wasn't stupid. The sky is filled with stars, she told Davey Stewart, her high school sweetheart. The Milky Way: the Wolf's Trail. But nobody responded, nobody even seemed to notice her. The bright air was laced with the spoor of caribou. She felt a sudden elbow in her back; she turned and awoke in a dark room, a stiff bed. I forgot the shopping today, she thought. There isn't any milk in the house, or any coffee.

Beside her in bed, the man slowly moved.

Sherryl sat up, her pupils gradually dilating. Eventually she discerned the motel room's clean, uncluttered angles. The thin and fragile dressing table, the water glasses wrapped in wax paper, the hot plate, the aluminum-foil hot cocoa packets.

'What's the matter, baby?' Andrew sat up beside her, his arm encircling her waist. 'Nightmare? Tell me, sweetheart. You can tell lover.' He kissed her neck, stroked her warm stomach.

'Please, Andy. Not now. Please.' Sherryl climbed out of bed. Her clothes lay folded on a wooden chair.

'Sorry. Forget it.' Andrew rolled over, adjusted his pillow and listened to the rustle of Sherryl's clothing.

Sherryl stood at the window, gazing out through the blinds. Stars and moon were occluded by a high haze of lamplight. She heard the distant hishing of street sweepers, and pulled on her blouse. Then she heard the rain begin, drumming hollowly against the cheap plywood door.

Andrew took his watch from the end table. The luminous dial said almost two a.m. 'I'll call you,' he said.

'No,' she said. 'I'll call you this time. I need a few days to think.' She opened the door and stepped out into the rain. They always do that, she thought. *They* have to be the ones who call, *they* have to be the ones who say when you'll meet or where you'll go. She pulled her coat collar up over her new perm, gripped the iron banister and descended one step at a time on darkling high heels. Puddles were already gathering on the warped cement stairs. 'It's as if we don't have any brains of our own,' she imagined herself telling Evelyn. 'And I'm sure that's just what they think. That we haven't got the brains we were born with. That we have to be told *everything*.' By the time she climbed into the Volvo the rain had ceased, as abruptly as if someone had thrown a switch. Her coat was soaked through, and she laid it out on the backseat to dry.

At this hour, the streets were practically deserted. She drove past a succession of shops and restaurants: Bob's Big Boy, Li'l Pickle Sandwiches, Al's Exotic Birds, Ralph's Market. Inside Long's Drugs empty aisles of hair supplies, pet food, household appliances and vitamin supplements were illuminated by pale, watery fluorescents, like the inside of an aquarium. 'It's not as if we couldn't do just as well without them,' she would continue, awaiting Evelyn's quick nods of agreement. 'I certainly didn't need to get married. I could have done just as well on my own. It's not as if it's some *man's* secret how to get by in this world. It's just a matter of keeping your feet on the ground, being objective about things, not fooling yourself. That's all there is to it. That's the big secret.'

As she turned on to Beverly Glen her high beams, sweeping through an alleyway, reflected off a pair of attentive red eyes. Being realistic, she thought, and heard the wolves emerge from alleyways, abandoned buildings, underground parking garages, their black callused paws pattering like rain against the damp streets. They loped alongside her car for short distances, trailed off to gobble stray snails and mice, paused to bite and scratch their fleas. She refused to look, driving on through the deserted city. The alternating traffic lights cast shifting patterns and colours across the glimmering asphalt, like rotating spotlights on aluminum Christmas trees. Wolves, men, lovers, cars, streets, cities, worlds, stars. The real and the unreal, the true and the untrue. Unless you're careful it all starts looking like a dream, it all seems pretty strange and impossible, she thought, while all across the city the wolves began to howl.

THE DARLING

Afterward Dolores Starr would lie on her bed with a sort of stunned and implicate amazement at the power of things, the power of that vast, soft universe of force contracting gently around her body like a hand. Dolores, she thought. Dolores, dolorous, dolorous star. She didn't feel hurt so much as bewildered and tired, as if she had awoken from a mere dream of struggle in some other, distant room filled with ballooning silence and white, intricate spaces. Usually by now Dad had returned to the kitchen to drink, but sometimes he took his gun from the clothes closet and waved it around for a while. 'Maybe we both learned a little lesson today, didn't we, Miss Teen Princess, little Miss Queen of the World.' Dad aimed his Walther P-38 at vanity mirror, cheesecloth curtains, Dolores's desktop crucifix. 'Ker-*pow*!' Dad said. 'Ker-pow, pow, *pow*! *That's* the only lesson most people ever learn, Miss Beautiful, Miss All the Boys Love Her. A bullet in the old brainpan, a crack on the head with a flat rock. Pow, bang. That's just about the only lasting truth *this* goddamn world's got to offer.' Dad's gun was very heavy and very solid, and filled the entire apartment with its weight and stress. Dolores liked to hold the gun in her hands too; the entire universe of force seemed to withdraw a little when Dolores took it from the closet; she felt as if she had more air to breathe. Most of all, though, she liked the sudden sound of it, and the way Dad looked at her as if she were someone strange and wholly unfamiliar to him. Then, very slowly, Dad lowered his head on to the kitchen table as Dolores moved his Jim Beam to one side. Dad's brains and blood virtually ruined the checkered tablecloth Dolores had bought at K Mart just that summer, and upstairs Mrs Morris struck the ceiling three times with her burnished mahogany cane. Mrs Morris was eighty-seven years old and lived alone. Mrs Morris lived on a pension, and had bad knees. Mrs Morris had raised four children of her own and often said she deserved a decent night's sleep every once in a blue moon.

She went to San Francisco and lied about her age, sat at a long Formica table littered with cigarette trays and ashes and solicited marketing surveys. All the operators wore miniature telephone headsets and resembled the crew of some shoddy spaceship. 'Have you graduated college within the last ten years?' Dolores asked people. 'Do you ever purchase Hallmark greeting cards? Do you have any children? House pets? Servants? Have you seen the recent television commercial for New Improved Wheatley Wheat Snaps? Have you ever been to Vermont?' She felt like a real adult now, with her own studio apartment on Fulton Street, a super-saver bus pass, a California Federal checking account and even a Versatellar cash card with her own secret code number. She developed a taste for Virginia Slims, piña coladas and Daniel, her Group Module Assistance Coordinator. Daniel was thirty-seven and lived in Brisbane. 'The abdominals – that's what goes first. The old midriff section. That's why I either run or swim every morning. That's why I do fifty gut crunchers every night.' Daniel had marvelous abdominals, a '67 Karmann Ghia convertible and a bookshelf filled with books. Dolores read Steinbeck's *The Grapes of Wrath*, Durrell's *The Alexandria Quartet* and Tolstoy's *The Death of Ivan Ilych* while Daniel jogged relentlessly down the peninsula, over San Bruno Mountain, around Candlestick Park. Dolores loved the world of books, which were a lot like adulthood, she thought. Both seemed rather smoothly improbable, at once perfectly real and perfectly contrived, like the uniform plaid tweed skirt and red wool sweater she had worn to a Catholic girls' school when she was very small. That was before Grandma died, and Dad started drinking.

Books made people different, she thought. That's why Daniel was different. That's why Dolores felt different every day, after every book. It felt as if every book she read somehow altered her chemical constitution. She thought she would be very happy with Daniel and his books, until the day he hit her. He hit her in the kitchen while she was washing up. He hit her because she hadn't been home when he called. He hit her because she just tried to tell him she was home all night. He hit her because he saw how other men looked at her and how she looked back. He hit her because he couldn't reach into that other part of her where she recognized other men. He hit her because he was just like Dad, he'd been fooling her all the time, he never really read all those books on the shelf. His face was red and damp and he'd been drinking with his friends at the ballpark, and three months later he thought she forgot what he did, thought the entire incident had gone far away when he crashed through the rear screen door, steaming with briny sweat in his Nike tank top and green nylon jogging trunks, and Dolores handed him his tall, cold protein shake. He took it down with one long parched swallow, his Adam's apple bobbing. The protein shake contained nonpasteurized whole milk, two fertile eggs, eight ounces of liquid protein, wheat germ, vitamin B complex and B12 and three heaping tablespoons of blue crystal Drano. It didn't kill him right away, though. He fell to the floor and pounded it, gurgling deeply in his chest and throat (ironically, Dolores thought, like bad plumbing) and pulled the telephone off the coffee table; it chimed brokenly. His mouth and eyes were pale and dry, and a hard green pellet popped from his throat and ricocheted off the blank, uncomprehending gaze of the Sony Visionstar. In a panic, Dolores sought razors in the bathroom, serrated knives in the kitchen, but discovered only Gillette Good News and disposable plastic cutlery. Finally she struck him twice on the back of the head with his simulation ivory and brass league-leading single average bowling trophy, spring 1982. His wallet held almost three hundred dollars cash, assorted credit and gas cards. She drove his Karmann Ghia convertible down Highway 5 to Los Angeles, and read Wilde's *The Picture of Dorian Gray* that night in the Van Nuys Motel 6. She liked *The Picture of Dorian Gray* very much.

She made large cash advances on all of Daniel's negotiable cards and opened a money-market liquid-assets account at the Sears Financial Network. She acquired a one-bedroom apartment in Fairfax, a clerical position at TRW and a 'new look' from Franklin and Schaeffer in West LA. Men often asked for her number and said complimentary things; men took her to expensive meals, nightclubs, sporting events. In her closet she gradually assembled entire wardrobes of memorabilia from the Dodgers, Raiders, Kings and Angels. Men were easy. They smiled, laughed, offered services, took checks.

They were grateful for the smallest attentions. Dolores carried a .380 automatic Beretta in her purse. She liked men, but that didn't mean she was going to take any chances.

Still, she felt vaguely dissatisfied with life. Something important seemed to be missing, or perhaps even beyond her comprehension. It was as if she were always forgetting something. She wanted to be happy. 'I guess it's because I never finished high school,' she told Michael one day at work. Michael sat with her at the Employee Benefits desk in Personnel. 'I guess I never figured out who I wanted to be, like maybe I've gone and wasted some special part of myself somewhere. Maybe because my mother left me when I was very small, I never felt very good about myself as a person. I know I go on lots of dates, but nobody seems to love me for who I really am.'

At Michael's suggestion she enrolled in night extension courses at Los Angeles City College. Every Tuesday and Thursday evening after work she attended lectures in abnormal psychology and functional human anatomy. Dr Peters, who taught functional human anatomy, looked just like Dad before he started drinking. He told her about the jugular, spine, meninges, bile duct. The body was just a delicate bubble, really, which could be broken open very easily; it made her nervous to contemplate twice each week her own physiochemical vulnerability. Infection, hemorrhages, renal failure, metastasis, stroke. Polio, eczema, muscular dystrophy, brain death. Every Friday in lab she dissected large cats and divulged complexes of lymph, nerve and muscle. Dolores much preferred Dr Deakin in her other class, where she tried to put out of her mind the dead cats with their rictus mouths and smell of formaldehyde. Dr Deakin was relatively young. He wore pressed and faded Levi's with white, tapered shirts and knit ties. He always punctuated his intense, Socratic monologues with profound, intriguing pauses. 'What does it mean . . . this word "abnormal"? And how do we know . . . when it truly applies?' He had an overgrown walrus mustache, and as he paced the lecture floor he gazed up into the high fluorescents as if entranced by gravid implications only he could see there, like some spiritual medium. 'Don't I think . . . *I'm* normal? And anytime you contradict me . . . don't I think *you're* abnormal? Don't we all like to define *ourselves* . . . as the "normal" ones?' Dolores quickly grew to love him. This was a man who understood the way the world worked; he could see far beyond himself into the eyes of other people, other people who hurt, cared, loved and cried. 'I certainly understand the importance of your class, no kidding,' she told him over a shared turkey-and-sprouts-on-rye at the corner Blimpies. 'I have had to deal with many abnormal people in my life, and I am just beginning to realize that they were not abnormal at all, but really were just normal, actually.'

Dr Deakin kept an immaculate duplex in Los Feliz, filled with lush hanging ferns and gleaming french windows, and Dolores cut his throat with one of the long steel carving knives from the immaculate and well-kept Spanish-style kitchen. He had been perfectly gentle and polite. She hadn't felt angry, or even perfunctory. There was just something in men which seemed to demand it now. Something in their eyes. It was like the look of seduction, really. The blood was suddenly everywhere, and if there was one thing Dolores was firmly resolved against from that night forward, it was knives. She began making a few strategic handgun investments. A .38 Special, a 9mm Parabellum. Dense compact Remington cartridges in a tidy cardboard box. She joined the National Rifle Association. She subscribed to *Guns and Ammo*.

Men were easy, but women were different. Women, in fact, were much more different from Dolores than men. Their glances click-clicked like the lenses of cameras, their tongues snapped faintly at you in reproach. They didn't like you talking to any of their men, and all the men in the world, it seemed, were their men. Women kept secrets, and liked to pay men special attentions in private. Dolores didn't even like women, though she hoped it was a condition which would change with maturity. Women practiced retributions on grand scales, they wielded sharp blades in profound ritual ceremonies beneath the earth in intricate, vast caverns filled with smoky incense and swelling female voices. Dolores never had a mother, so she never knew. Women shared a secret world of ritual, violence and redemption Dolores could only guess at.

'You know you gotta be careful in LA, don't you, Di?' Michael said, always bringing hot coffee in Styrofoam cups to her desk, candy bars, crackers. 'You read the papers, don't you? A single woman's got to be careful in this city; you know why? Because otherwise she'll get murdered, that's why. This city's filled with a lot of very crazy characters, Dolores. For example, just the other day I was reading about a whole club of murderers that lived out in the desert. The women, you see, would go to bars and pick up men. Then they'd take them out to the desert and they'd be murdered by the whole gang. It started out as an Indian cult, but then the white people started getting involved too. Even the white women. They skinned one man completely alive before they murdered him. So, what are you up to later, Di? Feel like a movie, maybe? Or dinner?'

They ate Thai, saw John Wayne in *Red River* and *Rio Bravo*. Dolores particularly liked Angie Dickinson, one of *Rio Bravo*'s costars who would go on later to star in the hit television series *Police Woman*. Angie Dickinson knew a woman could appear feminine and sexy and still know how to take care of herself. Michael sat quietly beside her and didn't even reach for her hand; she could see the movie flickering and inverted in his brown eyes. The movie theater was called the Vista and was located at the corner of Sunset and Hollywood. It was filled with misshapen shadows, stained and thinning velvet draperies, high abandoned balconies and enormous Egyptian-style statues, like some film festival in the Middle Kingdom. 'This used to be a gay theater,' Michael told her when they first sat down. He shifted uneasily in his seat. 'I can still smell them,' he said. 'Fucking queers.'

They bought ice cream next door and then drove to Griffith Park. Michael was silent, and Dolores felt a hard, cool pressure accumulating in her, like the thickness of gravity.

'What are you thinking?' Michael asked.

Every so often they passed the hunched figures of strange men in the shadows. Usually the strange men wore leather jackets; they had dark complexions and quick dark hands.

'I don't know. What are *you* thinking?'

Dolores unclasped her purse in her lap. Her right hand slid through the clutter of checkbooks, wadded Kleenex, random cosmetics and a dog-eared paperback copy of James M. Cain's *Mildred Pierce*, sensing the buried and unalterable weight of it there before she found it. It was always the same, she thought. Men who really loved you were filled with a sort of emptiness. Sometimes you wanted to fill that emptiness before it filled you. They pulled into a secluded parking lot in a grove of drooping jacarandas. Over the roof of the park the power lines hummed.

'It's not easy living alone in a town like LA,' Michael told her. 'I mean, it's not hard for someone like me, since I'm a highly independent person with a firm commitment to being exactly who I am. In fact, I can honestly say I have a very firm commitment to myself, which is not to sound egotistical or anything. It's just that I'm not one of those people, you know, who always needs someone telling them, like, this is who you are.' Michael reached under the driver's seat. 'Some people never understand,' he said.

Michael withdrew his .357 Magnum Desert Eagle just as Dolores withdrew her .380 Beretta Model 84, which featured a thirteen-round staggered magazine and a reversible release. A crumpled ball of Kleenex, dislodged from the trigger by Dolores's thumb, tumbled into her lap. It was a full moon outside that night, making only a dim impression against the high screen of smog.

Michael looked at Dolores's gun; then he looked at her eyes. He looked at her gun again. Finally he said, 'Don't you have trouble finding a good clean-burning handload for a piece like that?'

'I use Blue Dot,' Dolores said. 'I don't want to stress the barrel.'

They were married in July, bought a condo in the Valley and an Airedale pup named Bud. 'Bud's a pup who's going to have one solid family unit to depend on,' Michael said, dispatched a blizzard of résumés and acquired an administrative position at Lockheed in Burbank. 'You've got to believe in yourself if you're going to be happy in this life. Don't you, Bud? Don't you, fellah?' Michael scrubbed the Airedale's addled head between his fingers. The puppy gave a succinct yelp.

'Be careful,' Dolores said. 'You're hurting him.'

They went everywhere and did everything together. Tuesday evenings a self-actualization workshop in Sherman Oaks, Saturday afternoons an advanced gun care and safety program at the Van Nuys Police Academy. They installed a burglar alarm in their home, a doghouse in the blossoming yard and their mutual gun collection behind glass-paneled display cases in the den. 'It's like I have all the energy in the world now,' Michael said, and decided to build an arboretum in the backyard. 'It'll be like our summer home, a home away from home. We'll sit out there and drink iced tea all summer.' Michael loved their yard. 'Gardening is what I always needed,' he told her, returning from the nursery with marigolds, Lincoln roses, peat moss. 'It helps me make use of my more positive side, my life-affirming energy. I don't believe in anger anymore. I don't believe in hate. The world's got enough of those negative vibrations already without me making any more of them.' He installed floodlights on a high wooden vined terrace, and often worked on the yard alone and late into the evenings.

Dolores, meanwhile, would lie awake in bed at night and imagine the fluttery and somewhat appalled conspiracies of women. 'I'm not a thing- or a self-oriented person anymore,' Dolores told them. 'I'm a goal-oriented person now.' Deep in their immaculate caverns, the women murmured; they tried not to listen; they were deeply and mortally offended. 'I know you think I've just given in to some man, but that's not true. Michael isn't just some man. Michael respects me as a person. Michael respects me for being exactly who I am. You can't understand if you don't know that feeling how wonderful and important that feeling is. It's not something I can just explain.' Faint fibrillations, echoes, pulses. The women shared sonorous voices, impossible confidences, their hearts synchronously beating in the black caverns. Dolores didn't trust them; she wanted to get far away. Someday we'll have our own energy-sufficient cabin in the Pacific Northwest, she assured herself. We'll have trained Dobermans, electrified fences, canned goods. We'll have shortwave, and a proper armory which includes autoloading carbines and antiballistic missiles. Often she fell asleep before Michael came to bed, and when she awoke she could hear him already at work again in the backyard, striking the ground with spades, shovels, rakes; installing seeds, bulbs, determined little saplings. 'I thought we might have a little vegetable garden right here,' Michael told her. 'Then we don't have to worry so much about pesticides.' Dolores loved to stand at the large picture window and watch him work. Michael had long, fair-skinned hands which, finely etched with the brown dirt, resembled beautiful antique figurines recovered from some archaeological dig. 'I'm a high-energy sort of person,' he told her. 'I never sleep much.' Bud lay on the sunny grass and contemplated a hovering fly, his tiny body coiled like the spring of an HK P7. Weekends Dolores would sit on the faded-green lawn chair, drinking tall ice-cold drinks and smelling the moist upturned earth. Every few minutes Michael would look up from his work and smile at her. His tools lay about casually or leaned against the varnished pine fence like intimate friends at some large garden party, flaked with dirt like Michael's hands. There are places outside the world of men and women, Dolores thought. It's possible to live there safely and protected, like children with strong, enduring parents.

Then, one Sunday while Michael was pricing planters at Builders' Emporium, Bud uprooted the foot of a buried postman among a bed of Michael's blossoming dahlias. His shoelace was untied, and seemed to signify something, though Dolores was too shaken to decide exactly what. 'I felt impossibly alone,' she told Bud later, cradling him in her arms, dripping his still body with her tears. 'Everything I tried to believe was true about Michael was really just a lie. His honesty, fidelity, love – all lies. He never cared about me. He never wanted to share his life with me. He only wanted his own secret little world.' In the basement she had discovered jars of formaldehyde, handcuffs, ropes and enormous gray cloth sacks. 'He was never going to let me into that world. I was always going to be completely alone.' Bud was warm and motionless in her arms. It was dark out, and a full moon glowed faintly through the overcast. Then Dolores lay Bud in the trench in Michael's arms. She crossed Michael's arms across Bud, to keep him warm in the long darkness. Michael was wearing his three-piece Bill Blass double-breasted tweed, the same suit he wore the day they were married. Then, gently and with deep regret, Dolores distributed the damp brown earth across them both. It was as if she were burying herself in the tidy garden, placing her own humble body into the deep, whispering world of complicit women. The women themselves, though, weren't very happy. Nobody liked her there anymore. They didn't want her with them. Only men liked Dolores. Men and other men.

She drew the curtains on the picture window and every night she slept alone. The loneliness was immense and unsettling. She felt unpopulated black continents forming deep inside her body, jagged mossy peninsulas orbited by forlorn craggy islands and glimmering gray water. In the long evenings she sat beside the curtained picture window, motionless within a cone of light from the standing lamp like a display in some anthropological museum, feeling the hard relentless yearning of the planet underneath the yard, the secret articulations of graves and bodies. She never looked at the yard anymore, but only imagined it. Michael's abandoned tools just lay there gathering rust, their wooden handles cracked and splintering. The flower beds and vegetable garden would be overgrown with fast green weeds, the wheelbarrow overturned and covered with a thick gray impasto of cement. And Michael, of course, underneath all of it, still telling his lies, still lying to her all night and all day. She couldn't even hear the secret ceremonies of the women anymore. They had gone into deeper caverns where Dolores was no longer privileged. They were teaching her a little lesson. If she wanted to be Miss Little High and Mighty, if she wanted to be independent and on her own, then that's just what she'd have. Just herself; nobody else for her to feel any responsibility toward. Now all she could hear were the power lines buzzing on the high poles, the crickets wheezing, the dark planetary heart beating against the floors of her condo. Sometimes, particularly late at night after she had smoked too much marijuana and too many cigarettes, Michael would appear and attempt to comfort her in her darkest, loneliest hours. He would sit on the beige sofa, absently patting Bud's loose, volitionless head in his lap. 'You weren't secure enough in your individuality to allow me to be myself,' he told her. 'When people love each other, they have to trust each other as well, Di. I think you know that.'

Dolores never looked at him directly. She looked instead at the curtained window. She imagined bright spiders spinning their webs in the piles of moldering lumber Michael had purchased for the arboretum. 'I don't think I have anything left to say to you anymore,' she said.

Sometimes Michael moved to the faded-gray Barcalounger which Dolores had stitched together in places; sometimes the marijuana gave Dolores a vague sense of self-possession, as if she were in complete control of her own lungs, blood, heart. She could will her heart to slow down a bit; she could demand more oxygenation or less. Sometimes she felt as if she were sitting in another, blurred room far away from this one. Usually during these long waking dreams her mind returned to the same questions over and over again. She wondered if her mother was still alive somewhere. Would we recognize one another if we met unexpectedly on the street? she asked herself. Is there something chemical about the bond between a mother and her daughter, or are we just like any two strangers now? Maybe we'll become great friends by sheer chance someday. She will find my naïveté charming; she will teach me all about men. We'll go to movies together and take turns fixing dinner. We'll become devoted roommates, go to nightclubs, even dancing. In Europe women often go dancing together, and it doesn't necessarily mean they're lesbians or anything.

'You sit cooped up every night smoking grass,' Michael told her, the collapsed puppy draped across his knees like a hearth blanket. 'I think you've done enough feeling sorry for yourself to last a lifetime. I think it's time you took a little responsibility for your life, and stopped blaming everything on people you love.' Michael picked up the container of Herco smokeless shotgun powder from the coffee table. The shotgun, cleaned and loaded, was peering out from underneath Dolores's easy chair. 'You don't leave something like this sitting open all day long,' he said. 'It gets damp.' He affixed the aluminum lid with a quick hollow snap. 'Also, you better start looking in on the yard. The neighborhood cats have begun digging up Mrs Winslow again. If I were you, I'd go out there right now and check on Mrs Winslow.'

Dolores took the unfinished joint from the ashtray in her lap and lit it with her Cricket. A seed popped; a fragment of paper sparked and fluttered through the air. Without exhaling, Dolores asked, 'Who was Mrs Winslow?' Her eyes began to water.

Michael shrugged. In his lap, Bud's head rolled to one side, his large eyes dry and vacant like the eyes of some collapsed puppet. 'Just some lady worked at the library,' he said.

Then, one Friday evening in late summer, Dolores returned home from Von's to discover numerous police cars and ambulances parked in her driveway, their soft red and yellow emergency bulbs pulsing and spinning in the smoggy twilit air. They seemed vaguely sudden and incongruous, like emergency flares designating some roadside picnic. Dolores removed her groceries from the trunk, and a uniformed policeman at the door gazed at her with a sort of official complacency. Loaves of bread, a sack of red Delicious apples, gallons of distilled water in large clear-plastic jugs. Even though she lived alone, she liked to be prepared; if there was one thing life had taught her, you never knew what might happen next. She didn't feel surprised so much as slightly bemused when she was confronted by charges of multiple homicide with Birds Eye frozen vegetables under one arm, nachos and various snack crackers under the other. The arresting officer, Detective Rowlandson, was very kind. He asked her if the cuffs were comfortable. He transferred her frozen foods into the care of one of the random officers who were milling awkwardly about the small living room. The uncurtained window revealed a red, apocalyptic sunset and numerous men in white cloth shirts and trousers digging at the yard. Wearing surgical masks and gloves, they wrapped the moldering figures in white sheets and transferred them to stretchers which were then carried to the open chambers of patient white ambulances. When Detective Rowlandson drove her down the hill in his Eldorado the streets were filled with curious neighbors – housewives in faded terry-cloth robes, children leaning against their Stingray bicycles. 'Anything you say can and will he held against you in a court of law,' Detective Rowlandson told her, trying to find a classical station on the radio. 'I know,' Dolores said, 'and I think that's perfectly fair.' She was turning to look at the young officer in the backseat. The young officer was gazing aimlessly out the window. He seemed a little bored, or even homesick. When they arrived at the station Detective Rowlandson interrogated her in his private office, with another pair of uniformed patrolmen at the door and a cassette tape recorder whirring on the desktop. 'Maybe you'd like a little soda or something?' he asked. 'Maybe your throat's getting a little parched?' They were all very kind, Dolores thought. Even when they don't really know what's going on, men really do try to do their best. Men really do care about the unapproachable world of women.

She was awarded a private cell and instant, irremediable celebrity. 'I can't say I'm proud of what I've done,' she told the media, which were assembled around her in a bright fluorescent room of flashing cameras and buzzing tape machines. The journalists sat poised on the edge of their aluminum chairs as if expecting some race to commence without a second's notice. 'It's not like I'm stupid either, since I always did well in school whenever I bothered to apply myself, and Dr Weinstein, who is one of the very kind doctors visiting me while I am incarcerated, says I performed exceptionally well on the Wechsler Adult Intelligence Scale. I guess I can only blame my poor upbringing, being as that my mother left me when I was very little, and as my father beat me when I was little and took advantage of me in many ways which are too delicate to be gotten into at this time and place. But anyway, I can't blame everything on my parents, since I am a grown-up woman who must take responsibility for herself, and so I would like to say that I am solely responsible for all those dead bodies buried in my yard' – which initiated a blizzard of bursting flashbulbs – 'and of course for my good husband Michael's senseless and untimely demise as well, and if I get sent to the electric chair I will certainly deserve every minute of it since Michael was the kindest, most loving husband the world has ever known, and he was certainly the only man who ever actually tried to understand and care for me in a totally unselfish and caring manner. Thank you very much.'

*

Dolores's private cell was in the women's maximum-security prison in Lancaster. She had a toilet, a washbasin and a prison-issue towel, soap and toothbrush. She had a rough green khaki blanket and bristly sheets. Every afternoon they took her out alone to exercise in the courtyard. She walked calmly around the painted white basketball tableau. She did sit-ups and leg lifts, pausing occasionally to gaze up at the bright California sky. The guards were all women. When she saw the other inmates, they were all women. They all had hard, coarse expressions. Sometimes, far off down the distant cement corridors, Dolores could hear a young woman crying. She sounded very young, almost a child.

Dolores was entering her Russian novel phase. She read *Crime and Punishment, Anna Karenina* and *War and Peace.* For the first time in her life, Dolores felt at peace with herself and her innermost being. *It's like I never had a chance before to actually understand what it was like to be totally on my own*, she wrote on her pad of white paper, which was inspected every evening by one of the uniformed guards. *Maybe if I had only had a chance to get to know myself without other people around me all the time making me feel like somebody I wasn't, I wouldn't have killed all those nice people.* She contemplated writing her own autobiography and publishing it under the title *Bad Love.* Her cell was absolutely silent for hours at a time. In fact, Dolores rarely saw any men at all. She felt denser, more compact and more real. It was as if her entire body were filling up with sand. She refused newspapers and magazines. She was a quiet, respectable hermit living alone in a deep cave. She was contemplating convoluted and transcendent things. *Some things you just can't explain*, she told her writing pad. *Sometimes too you can be just happy not explaining them either.*

'They've got you now, baby,' Michael said, picking at the celery on her evening meal tray. 'As they say in the movies – the jig is up.'

'We'll see,' Dolores said. She felt a vague glimmer of hope, one which filled her with impossible sadness.

A few days later Dr Weinstein fell in love with her, and she knew all the peace she had finally grown accustomed to would not last. 'Primitive man didn't draw pictures on his walls because he liked pretty *pictures*, for chrissakes,' Dr Weinstein told her during one of his visits, trying to act like he wasn't in love with her, like he was different from other men. 'It's not like *Neanderthalus australopithecus* buried his dead out of fucking sympathy and compassion. How much sympathy and compassion do you think you'd get from a *Neanderthalus australo-*

pithecus? Not too damn much, that's how much. Not too damn much at all.' He showed her a picture of *Neanderthalus australopithecus* in a large library edition of *The Encyclopedia of Human Anthropology*. 'You see that guy? You see those teeth, that brow? Why do you think he painted pictures on the wall? For the same reason he ate the still-bloody hearts of the rival tribesmen he killed, that's why. He was appropriating the soul and strength of significant others. Family, beasts, enemies. The sun and the fucking moon, that's what.'

He carried a black leather briefcase. He wore a dark suit and glasses. At first he appeared only every few weeks or so and asked her to complete psychological profiles, write personal compositions and analyze photographs of men, women and children in family situations. Then he began arriving every afternoon just as the lunch trays were being collected by a trusty on a wobbly aluminum cart. Sometimes he talked for hours while Dolores sat on her cot, her hands folded between her knees, her blank gaze trained upon the concrete floor which she had scrubbed clean just that morning.

'We do it every day,' Dr Weinstein said. He held the briefcase in his lap with his left hand; his right hand gestured vacantly at the cold and empty air. 'We appropriate the souls and strengths of other people. It's just that most of us don't have to kill them, babe. You know what I'm saying at you, Di? You don't mind if I call you Di, do you? I saw it on your Wechsler examination under preferred nick-names.' He offered her cigarettes and she smoked them, inhaling the grainy, desultory smoke, watching it settle across the stone floors like morning mist in a swamp. 'Love and aggression are the same thing in human society. They're both responses to the same biochemical hums and pops. You love or hate the other and you want to blast them. You want to break them down into their elements and swallow them. You want to make them one with yourself by devouring, feasting, obliterating. Then they're part of you, aren't they, babe? Then *you're* in complete control. It's a biochemical desire, but when we live in society, see, we learn to develop displacements for those desires. We learn to turn acts into symbolic intentions. You don't *do*, in other words, Di. You learn to seem *not* to do, if you know what I mean. But you really *do* do, secretly, but only in your mind. Only you, babe, you don't know how to do that. You think there's just your mind and the world, that the world's the only object your mind's got to act on. You have to learn to invent other objects. You have to learn to compensate for your desires by instituting certain ritual behaviors in your seriously addled and definitely very sociopathic psyche, Di – and I think I can say that much for certain. Definitely sociopathic. These are things you're supposed to learn – when you're raised properly. But you haven't been raised properly. You've got to be raised all over, right? You see what I'm saying, Di? You've got to be raised all over again.'

Dr Weinstein testified at her first court appearance, and the charges were dropped on grounds of insanity. 'Let's say they were pretty firm grounds, Di. Let's say we had a fucking continent full of firm ground for that one,' Dr Weinstein told her in the government car that took them to a county holding cell after the hearing. Three days later Dolores was remanded to the custody of the state psychiatric clinic in Reseda and, three months after that, quietly transferred to Dr Weinstein's private facilities in Napa County. It was a different place from prison, and Dolores didn't like it. The grounds were green and unenclosed, with a view of rolling hills patched with vineyards. Dolores was apportioned her own private room, wardrobe, library and lawn chair. The patients here were all very quiet and composed, and didn't look disturbed at all that Dolores could tell. Rachel, an attractive, fortyish redhead, told her, 'When my husband closed down our savings account and ran off with his secretary to Buenos Aires, I guess I just couldn't cope.' Rachel was wearing a polka-dot cotton summer dress and reading *Cosmopolitan*.

Dr Weinsteien was personally committed to raising Dolores all over again. Her diet was strictly regulated. Listlessly, she attended the clinic's mandatory Exercycle workouts. Her blood pressure was intently monitored, her saliva, feces and urine; two interns from UCLA Medical Center received a grant to monitor her endorphins. She was steeped in megavitamins and zinc; she suffered a high colonic. 'Symbolic displacement,' Dr Weinstein told her after each morning's 'contact therapy' interview in his office. 'There are certain amine molecules manufactured in the adrenal gland which generate rage. There's good rage and there's bad rage, and your rage, Di, is very bad. These amines are then conditioned and modified by those massive discharges of the endocrine system concerned with reproduction. Reproduction is something your body anticipates around the clock; your body's always preparing you for reproduction, Di.' He took her hand and commented on her long, strong fingers; then he brushed a vein with alcohol and inserted the needle. 'It's at the confluence of rage and sex where we're trying to get,' he said. 'We're trying to draw a line between intentionality and action, pure rage and sudden sex. That's the line that's been eliminated in you, babe. We're going to replace it. We're going to draw it fast and hard.' She received the injection three times a day, and Dr Weinstein began taking her on what he liked to call 'field studies.' They drove to Marin County and purchased a new Volvo. They went shopping for clothes, curtains, sheets, dishes. Dolores had never really enjoyed shopping that much before, but now she craved it like potato chips; it took her away from herself; she could lose herself in the vast chattering communities of women. Afterward she and Dr Weinstein would return to his private office at the clinic and watch television; often they attended movies and plays together. He pronounced her fit for the home-based phase of her therapy. They were married in August and set up housekeeping in a beautiful, isolated two-storey country house in Sebastopol. Dolores worked mornings at the local day-care center while Dr Weinstein was at the clinic. Then she had the rest of the day to watch television. She didn't like books anymore. Dr Weinstein's Literary Guild selections gazed down mutely from the high mahogany bookshelves like zoological specimens cradled in formaldehyde jars.

She still thought of murdering him. Not every day, but periodically. At these times she felt herself inflating with a strange, unidentifiable sensation. Her heart began to pound; the backs of her hands began to itch. Her face grew flushed and hot, and she developed splitting migraines. She had never felt so intensely aware of the flux and convection of her own blood before. 'You're learning, Di. You're learning to accept the limitations of your own body, your own mind.' Dr Weinstein sat in the stuffed chair beside the jetting blue flames of the gas fireplace. The latest issue of *The American Journal of Psychiatric Medicine* was propped open against his knee. 'Fix us a cup of coffee, babe. Sit down and relax with me.' Dolores went into the kitchen and saw the immaculate wooden cooking utensils hanging from the varnished redwood cabinets. Then she went out the back door and made the screen door slam. She drove their second car to the Emporium mall and had a Bloody Mary at Marie Calendar's. She was still filling up with the unidentifiable feelings. She tried to repress them, but she didn't know what she was repressing. Terrible anger and rage, she suspected. That was what Dr Weinstein told her; that's what the daily injections were investing her with. She was frightened and disoriented. She sat down at a row of plastic stools near a wide mirrored fountain. Blue water streamed from the blowholes of glass dolphins. The fear grew more and more terrible as she watched the pulsing crowds and families. Teenage girls emblazoned with cosmetics. Young couples pushing dazed babies in carriages with tiny stuffed toys dangling from their fabric awnings. Packs of young men with faces flushed from marijuana. It wasn't fear anymore, it was panic. Dolores felt panicked but she couldn't move; she couldn't face the crowds of people; she couldn't face the acres of cars in the vast parking lots. She started to cry and cry. She had never cried in front of strangers before. When someone tried to touch her she pulled away and screamed at them. She didn't know what she screamed, but she knew she didn't want anybody near her. She just wanted to cry and cry, as if the entire world had ended and now only its unaccountable sadness was left, filling her and filling her like the hard colorless rage with which she desperately desired to murder Dr Weinstein.

After these 'episodes' Dolores would be sedated and kept overnight at the clinic. In the morning, Dr Weinstein would drive her home in the Volvo, usually playing Philip Glass on the car stereo. 'It takes a while to adjust,' he told her. 'We're teaching your entire body how to behave all over again. We're teaching it how to feel and breathe.' His right hand reached out and held hers in her lap. She felt enervated and thick with barbiturates. Outside, the entire landscape was blurred and indistinct. 'We're teaching you how to love, babe. We're teaching you how to love without hurting anybody.' Dolores began to feel extraordinarily lonely and weak. 'And you know I love you, Di. You know that, don't you?'

She couldn't even remember the faces of any of her old lovers anymore. Their memory seemed to be draining easily from her like water from a tub. She could remember their names – Daniel, Dr Deakin, Michael, Dad – but she couldn't remember anything about the quality of their presence, the fabric of their skin or voice or hair, the strength of their muscles or intestines. In the long summer afternoons she would just sit outside in the sculpted front garden, wearing her cashmere robe, black stockings and a silk teddy, beside an ice chest filled with margaritas on the wrought-iron lawn table next to the Valium prescription and her strewn cosmetics, and gaze aimlessly at the blue sky, green trees and topiary hedges. There was just a dark inchoate sadness now, formless and buzzing. 'It's the recognition that you're alone, babe. It's the human condition, it just means you're sane, that's all. It means you're not swallowing people. It means you know who you are, and who they are, and that line where the twain shall not meet. You're developing a nice clean, bright soul now, like Billy's bright T-shirt in a television detergent commercial. You've got your own world inside now, babe. You're ready to live your own life.' Dolores sipped her margarita and thought about *Neanderthal australopithecus*'s cave. Someone had expunged all the pale etchings of bison and mammoth from the rough basalt walls. There was nobody left in the cave at all anymore, not even the flickering fire or the smell of roasting meat. Dolores lit a cigarette and looked at the impossibly blue sky. For a moment she thought she might start crying, but then she didn't.

The following summer Dr Weinstein pronounced her cured, and exactly one year after that she gave birth to a nine-pound baby boy. The baby had a full head of black matted hair when he was presented to her by the nurse; his eyes were squeezed shut with pain and screaming. She held him against her breasts and listened to his heart beating in her private room while Dr Weinstein sat beside her, beaming like a streetlamp and holding her hand. After a few days of bloodless discussion, they named the baby Andrew, in honor of Dolores's dad.

SWEET LADIES, GOOD NIGHT, GOOD NIGHT

'Women only seem like nature when nature's not around. But women are more deliberate than that. A woman's love is completely artificial, just like her cosmetics. Woman's love has design and strategy in it, unlike man's. I think men have been getting bad press for a few thousand years now, or ever since some woman elicited poetry.' Arnold Bromley was picking at his right ear with a twisted paper match and sitting at the counter in Coco's on MacArthur Boulevard. Empty saucers and a single coffee cup sat before him on the counter. One saucer was tracked with chocolate cake topping, the other with coffee and cream.

'Did you want more coffee?' the waitress asked. Her name was Michelle. Michelle was an undergraduate in premed at the university.

Arnold looked at her a minute. She was thin, pretty and immensely uninterested. 'I guess I'll just have the check,' he said.

Then Arnold drove to Estelle's place and found Estelle in bed with a man whose name, he learned later, was Bob Reilley. Estelle, as it turned out, had met Bob Reilley at a neighborhood Safeway approximately two days before.

'There's something I've been meaning to tell you,' Estelle said. Bob Reilley had pulled a towel over himself and, out of courtesy, was putting out his recently ignited cigarette. 'I don't think I've been entirely fair with you lately, Arnold,' she said. 'Frankly, I think you can find a woman who'll treat you a lot better than I do.' Bob Reilley yawned distantly, gazing at his extinguished cigarette in the glass ashtray beside the bed.

Arnold let himself out of the front door and pushed Estelle's keys through the mailbox grate. He got in his still-warm car and drove around for a few hours. This time, he stopped for coffee at Denny's.

He looked at the menu just long enough to give himself time to unwind. He already knew the menu by heart.

'Chocolate cake and coffee,' he said after a while.

The tiny plastic nametag on the waitress's blouse said Mary-Anne.

'Sex is only more acute than love. But that doesn't mean love is any less real,' Arnold said.

Her name was Jilly, and she worked at I. Magnin. She wore sheer skirts and nylons and bright, almost fluorescent makeup. She had beautiful legs.

'Love's the way we learn the world's immediacy, and so we need it more. Love holds things in context. Love's the melody, and sex just the individual chords in a song. That's why love's so much more important to us when we're alone. Sometimes I think I'd give up my entire life for love. Sometimes I think I already have. Would you like to stop here? Would you like an ice cream?' They were standing outside Häagen-Dazs in the enormous shopping mall. The mall was filled with couples, families and running children. The windows displayed elegant clothes, furniture, table settings and expensive toys. The air reverberated with golden light. Plastic stools and benches down the centers of each aisle were crowded with dazed, expressionless shoppers and their baggage. The shoppers seemed to be contemplating nothing. They seemed very enervated, and annoyed with themselves for feeling so tired.

'How about a chocolate chocolate-chip ice cream?' Arnold said. 'Does that sound good? Did I tell you, Jilly? Did I tell you you have beautiful legs?'

Jilly had a boyfriend named Bob. 'But you're a lot more interesting than Bob,' Jilly said. 'You're always saying interesting things. It's not easy for a person to feel very intelligent about themselves when Bob's around. All Bob ever talks about is his car. This is what he's going to do to his car. This is what he's bought for his car. It's like living with instant replay. It's like living with Bob's car. Sometimes it's like not living with anybody at all.'

Jilly broke up with Bob, and Arnold knew he was on warning. From that day forward Arnold tried to be as interesting as possible. He took

Jilly to movies, museums, sporting events, RV and boat shows, carnivals and circuses. He told her about politics, language, culture, literature, theater, war, domestic surveillance, wok cooking, the history of unions, homemade salad dressings and just about anything else he could make up. But then one night at her apartment he just wanted to watch television. He couldn't think of anything interesting to say and, for that one night, he didn't really feel like saying anything interesting, either.

Jilly broke up with Arnold the following weekend and went back to Bob. 'I'm a one-man woman, Arnold,' she said. 'And I'm afraid Bob's that man. It's not that I don't still care about you a lot. It's just I think you'll be better off finding the sort of woman that's right for you, and I'm afraid that woman's just not me. Now here, I've packed all your things in this bag, so if you could take just a quick look around. I'm expecting Bob here any minute. If you see his truck, try to step on it, will you? You'll know it's Bob's truck if it says "Bob Reilley's Plumbing Supplies" on the side.'

Arnold purchased new shirts and after-shave. He purchased new shoes. He restyled his hair and purchased a blow dryer. He even purchased a new futon with a straw mat base, hand-painted sheets and matching pillowcases. When he got home and looked at himself in the mirror, he liked what he saw. He looked different and he smelled different. He looked like a man who knew what he wanted from life, and was prepared to go out and get it. He smelled like a man who wouldn't take no for an answer.

'I guess I've always been an other-directed sort of person,' Lydia Sanchez said, eating mushroom linguini at Marie Calendar's in El Toro. 'A lot of people are self-oriented sorts of persons, because they've never progressed beyond the nipple stage. That doesn't mean I'll let other people take advantage of me. And *that* doesn't mean I'm just thinking about myself because I won't let other people take advantage of me. When you let other people take advantage of you, you're doing them a severe disservice. You're not letting them learn what it's like to live in a real world where people don't let you take advantage of them all the time. I received an EST summer scholarship when I was eighteen years old. I was audited by scientologists and believe it was the most profound, unforgettable experience of my entire life. Now I like to think of myself as a sort of Rosicrucian

Buddhist, and I'm presently attending night classes at Valley College to receive my BA in cultural anthropology. During the week I'm a receptionist for Dr Brady, Credit Dentist. You know Dr Brady, don't you? He has those awful commercials on TV. In reality, though, Dr Brady happens to be a really nice man.'

Arnold looked over both shoulders, trying to locate the waitress who had virtually disappeared with his order of gin and tonic. Marie Calendar's was filled with young, attractive and well-groomed couples, all very much like Arnold and Lydia, who shamelessly flirted not only with one another but with members of the many other attractive couples seated at tables around them.

As if in afterthought, Lydia said, 'I really love eating at Marie Calendar's.'

Arnold loved Lydia because she did to him in bed everything he could possibly hope for and didn't even wait for him to ask. They had many wonderful times together for almost ten days. Then, like rain, it happened. One night Arnold brought the makings for sandwiches over to Lydia's apartment and they watched television together. 'You know, this is all anyone ever needs,' Arnold told her, gesturing with his pastrami salad on French roll. 'I mean, we've got decent food, we're in love and there's a good movie on TV. It's not like we need those things other people need to be happy. You know, like fancy restaurants, or going to big stage shows. You know, like *Cats* or something. It just goes to show you don't need a lot of money or anything just to be happy.'

That night, after Arnold had gone home alone, he felt it break in the dark air. It wasn't a sound so much as a beat, like a sudden shift in temperature, or some seismic tremor deep underground. Arnold sat up in bed suddenly. He looked around his dark room at the lopsided bureau, the paperbacks on their shelves and the tiny Casio electronic alarm clock on the orange crate beside his bed. There was something reproachful about it. The time was three forty-five a.m.

Lydia stopped returning his calls. 'Is your machine working all right?' Arnold asked her answering machine. 'Or are you just not getting my messages?' Finally, early the following month, Arnold went to Hamburger Hamlet and had three quick gin and tonics on an empty stomach. The gin filled him with resolve, fortitude and self-confidence. 'Call it arrogance if you want,' Arnold told himself, getting back into his car. 'That's fine by me. You've just got to be a little arrogant in this life if you're ever going to amount to anything.' He parked outside Lydia's office on Van Nuys Boulevard and waited for her to come out.

'I want you back in my life,' Arnold said when she appeared. She was accompanied by two young women who wore dental assistant uniforms. They both smiled at each other over Lydia's shoulder, as if they had been anticipating Arnold's sudden courage all along. 'I want us to spend the rest of our lives together,' Arnold said.

Lydia removed the bolus of pink gum from her mouth and placed it delicately on the back of a wooden bus bench. Her eyes were deep, black and remorseless. 'No,' Lydia said. 'No, Arnold. I'm afraid I don't think that would be a very good idea at all.'

His name was Bob, she told Arnold finally in a letter. Bob Reilley. She and Bob were very much in love. But that didn't mean she wasn't still concerned for Arnold's own happiness and peace of mind. She sincerely hoped that, someday, after they had both, as she put it, 'straightened out their lives emotionally and gotten to know themselves deep inside,' they could be friends again. Perhaps, at some elaborate reconciliatory dinner, he could meet Bob face-to-face, and she could meet his new girlfriend, and they would all feel very, very happy things had turned out so very well for everyone after all.

Arnold began putting on weight. He drank Budweiser by the case, and spent all his free time in front of his cable television, dialing the remote control and watching baseball games. He began checking books out of the library concerning low blood sugar and dietary disorders. Then, suddenly one morning, he noticed his hairline was receding. For the first time in his life he developed faint rashes of acne around his jaw and neck, particularly after he shaved. He felt generally tired and listless. His apartment seemed to be growing tinier and more unkempt. The new hand-painted sheets on his futon even began to unravel. Every few weeks or so he went to the supermarket filled with predictable resolve. He purchased fresh vegetables, low carbohydrate oils and salad dressings. When he returned home, however, he already lacked sufficient energy for mixing them together. Over succeeding days the carrots and celery shriveled and diminished in the refrigerator. They began to resemble shrunken heads and genitalia, or the archipelagoes of fantasy kingdoms in garish paperback books. Sometimes they mildewed and he threw them in the trash.

'Love's an illusion we can't always afford,' he said at Sambo's. The half-eaten chocolate cake was turning stale on his plate. The waitress, whose name was Polly, brought him more decaf. 'Sometimes love isn't love at all. Sometimes it's just the past. Sometimes it's just some moment we never really knew.' Arnold grew silent for long periods of time, watching the steam unravel from his coffee. There always seemed to be something important he meant to say, but he couldn't recall exactly what that something might be.

'I live right down the street,' Polly said when he brought his bill to the register. Polly had bright, flat white teeth. Polly was in the university drama department. 'Why don't you give me a call sometime?' Polly said.

He took long walks now rather than drove. At work, many of the female employees began showing him obvious and elaborate concern.

'Something wrong? Something wrong at home, Arnold? Bad financial investments? Vitamin deficiency? Have you ever heard of Epstein-Barr virus syndrome? Do you know a lot of young men are diagnosed with Epstein-Barr virus syndrome these days? It's mainly dietary, Arnold. It means there's something seriously wrong with your diet.' Sometimes one of the women's long, brightly lacquered hands would flash and lightly brush Arnold's warm forehead. When they came up close to him he could smell their moist perfume. A silk sleeve might brush his bare arm. He might catch a glimpse of cleavage.

'You need someone who'll take care of you,' they said. 'You need someone who loves you. Why don't you lie down for a minute, Arnold? Why don't you come over tonight and I'll fix you a nice, hot dinner?'

Arnold, suddenly out of breath, would walk quickly from the office and into the men's room. There, among the harsh fluorescents (which felt to his skin more like texture than light) he would gaze at his florid, spotty face in the bright mirror, and feel the heart pounding in his chest like something mechanical and estranged. His stomach and colon fluttered. That spring he began to develop a bleeding duodenal ulcer. Whenever he returned home, the message light on his answering machine was flashing.

'Hi, Arnold? This is Vicky. Do you remember me? Do you remember San Diego? I was just thinking about you. I mean, if you wanted to give me a call or anything, I'll be at home. Okay, Arnold? And if you don't want to give me a call, well, that's all right too . . .'

*

Bob Reilley's plumbing supplies franchise was located in Tustin, and one smoggy, humid day in April Arnold drove out to his office carrying in his right-hand jacket pocket a Smith Model 686 .357 Magnum which he had purchased earlier that month at Grant's for Guns in Costa Mesa. The gun felt very heavy in his pocket. It was a clean, comfortable sort of weight, however. There was a lot of security in the weight of a .357 Magnum, Arnold Bromley thought.

'It's not like I haven't been through all this before,' Bob Reilley said. 'It's not like I don't understand exactly how you feel, or why you've aimed all your misdirected frustration at me.' Bob Reilley was leaning back in his padded leather swivel chair. His feet were propped up on the desk, and his hands folded behind his head. Bob Reilley had changed a lot since that evening more than two years ago when Arnold had discovered him in Estelle Constantine's bed. Bob Reilley had developed a very significant belly, flaky red hands and face and a large bald patch across which he had combed elaborate swirls of greasy, dyed-black hair. 'I do sympathize with you, Arnold. I really do.' Arnold thought that Bob Reilly was just about the most nondescript-looking man he had ever seen in his entire life.

Arnold took a seat in the unsprung sofa facing Bob Reilley's desk. The sofa was inadequately covered by a thinning corduroy blanket. Arnold had let himself in through the unlatched back door of the abandoned plumbing supplies office, where the long aisles were filled with dismantled shelves, refuse and long, intricate ropes of dust. Only one, semicharred fluorescent bulb flickered overhead. The door marked MANAGER had been open when Arnold walked in and found Bob Reilley asleep at his desk.

Without removing his feet from the desk, Bob Reilley reached behind himself for one of two half-empty bottles of Jim Beam and a glass from the bookshelf. 'It always hurts to know we're not the only one. It always hurts to learn love isn't something that's been waiting for us, but just something that happened to be around when we got there. Look, if it's any compensation, I've been through it myself, you know. I mean – excuse me. Would you like a drink?'

Arnold looked thoughtfully at the proferred Jim Beam. He held the gun in both hands, the handle leaning against his crossed right knee. 'No, thank you,' he said.

'Well, don't want to let it go to waste.' Bob Reilley finished his glass and poured another, taller one. 'As I was saying – I've been through it myself, Arnold. It's not like I'm some heartbreaker or something. It's not like I just run around Southern California trying to make time with other men's women.' Arnold wiped his nose with the back of his hand. Then, thoughtfully, he wiped the back of his hand against the thigh of his gray slacks. 'Look, we're both single guys, right? Every woman we meet's just *bound* to belong to *some* damn guy or another, right? Women aren't like us, Arnie. They don't sit around brooding very long when things are over with some guy. They move on. So what are guys like you and I supposed to do? Keep our eyes on the obits or something, like apartment hunters in New York? The race goes to the swift, Arnie. The quick and fleet of foot. Don't tell me you've never done it. Don't you start playing holier-than-thou with that stupid gun.'

'This isn't personal,' Arnold Bromley said, indicating his heavy gun with a breezy gesture. 'This one's for abstractions. This one's for love.'

'Oh my. Sorry, Arnie.' Bob Reilley had spluttered Jim Beam all over the back of his arm. 'Love. I get you. We're talking love now, are we?'

Bob Reilley wedged his glass between his thighs and poured it full again, splashing some into his lap. The abandoned offices echoed with the buzz and flutter of the broken fluorescent. Bob Reilley's voice seemed to increase suddenly in both volume and confidence. 'That's what women are for, are they, Arnie? Love. Love and beauty, right? You tell me about it, Arnie. You see those fucking empty shelves out there? What do you think those empty shelves used to contain? Well, I'll tell you what they used to contain. They used to contain the most complete hardware maintenance equipment selection in the tricounty area. That's what love does to a guy. It takes a big successful business and turns it into nothing. There was one girl, Arnie, she broke up with me, right? For six months she phoned me ten or twelve times every night and called me a bastard for letting her break up with me. She showed up on my porch with a hand ax. I took it away and the police arrested me for asault. Women send their boyfriends after me all the time. They want to teach me lessons because I didn't love them enough. They want to teach me lessons because I loved them too much. I've had my telephone number changed about a hundred times. Do you want to know how many women are out there, Arnie? How many women in the world there are who hate my guts? I haven't any idea, Arnie. I really don't. Women aren't action, Arnie. They're judgment. And I think judgment's a lot worse than action just about any day.'

Bob Reilley took another long drink and wiped his mouth. 'So what are you going to do with that gun, Arnie? Shoot me? Put a bullet in my head? I'm really scared. I go to bankruptcy court on Friday. I've been sued, assaulted, defamed, perjured, disavowed, and once, I was once almost even impaled. How do you think things'll turn out in bankruptcy court this Friday, Arnie? The judge's a woman. So's my lawyer. Women are everywhere, Arnie. It's not like the good old days. Today there are women in government and banking. There are women in distributorships, construction, and even controlling the entertainment industry. My broker was a woman. We went out two weeks or so. I lost seventy-three thousand dollars after that one was over, Arnie – seventy-three thousand. Then, in February, one of them burned down my house. It was a nice house. The insurance policy had lapsed. No, my insurance agent wasn't a woman. But my insurance agent's wife was. My insurance agent's wife was probably the biggest mistake of them all. Two nights in a Best Western in La Jolla. If I could go back in my life and change just one thing, it would probably be those two nights with my insurance agent's wife in La Jolla.'

'You're right,' Arnold said, holding the gun just as firmly. 'Women aren't nature. Women are judgment. Women are language.'

'It's like there's nothing we can do, Arnie. We're damned if we do and damned if we don't. It's like no matter what we do or how we behave, they'll still be able to say anything they want about why we did it. What's our behavior matter if they can just remake behaviour's rules anytime they want?'

'It's a losing battle.'

'It's like time.'

'It's like we're always being betrayed by our own bodies.'

'It's women, Arnie. It's not us. It's not you and I, pal.' Bob Reilley got up fom his desk. 'We got to stand up for one another, right, Arnie? We got to stick together.' He was walking slowly, confidently across the room. His right hand reached out for Arnold Bromley's gun. 'It's ridiculous for *us* to be arguing, isn't it? It's language again. It's like we're playing right into their hands.'

Arnold Bromley stood up too. 'No, Bob. This isn't language. This is reality. And now, if you don't mind, I think I'll get back to language again.' At that point, Arnold shot Bob Reilley six times in the back of the head as he made his break for the door.

In the storage aisle, the fluorescent extinguished with a sudden audible pop. Everything went completely black.

'Good night, sweetheart,' Arnold said. 'Good night, Bob.'

THE WIND BOX

Sometimes he stood baffled by white sunlight and just watched cars go by. Bright convertibles, battered Fords and Chevrolets, roaring buses, long smooth white limousines with dark windows. The heat seemed to pull at him, like the sunlight flashing off chrome and windshields, like his sense of great quakes and tremblers stirring dreamlessly awake in the deep earth beneath his feet. Obdurate mountains grew down there, massive and subcrustal seas and continents. That was the first thing he noticed about Southern California: the sea of movement embedded in it. For the first few days he slept in his car on Mulholland Drive, but eventually he found a moldering bungalow studio apartment off Vineland Avenue in Burbank, and a large unsprung stuffed chair in an abandoned lot nearby. He purchased a curiously stained mattress from the local Salvation Army and transported it home in a neglected shopping cart. Large, dense men with rough, menacing beards, sun-stained leather jackets and reflecting mirrored sunglasses intensely disregarded him as they kicked their motorcycle ignitions with spurred black boots. Sometimes at night he would sit on his front porch in his tattered green lawn chair and just watch the high, buzzing power lines. He could hear the power lines when he slept; they invested everything, even his dreams, with a cool staticky mist. It was the only coolness he knew that summer.

The heat woke him every morning at six a.m. He fixed instant coffee, Cocoa Puffs, Potato Stix with ketchup, hearing the blare of televisions converge in the littered courtyard. The screeching of cartoon automobiles, the abrupt explosions of animated cats, a deep-throated, impossibly voiced villain crying again and again, 'Destroy the transformers! Destroy the transformers!' He sat at his wobbly table in the muggy kitchenette, gazing out the window at refuse in the courtyard and mangy, addled cats dozing underneath a tireless Ford Falcon. Unraveling truck retreads lay strewn about, fragments of irreparable furniture, unidentifiable hunks of metal, a sun-bleached stray number 12 billiard ball, and everywhere thousands of crushed and decomposing cigarette butts like gravel at the bottom of a tropical fish tank. There were times when it seemed to him the entire world had ended. There were other times when he just wished it would.

Vast rolling geothermal plates slipped and heaved, buildings toppled, freeway off-ramps collapsed. Earthquakes crashed through the valley like benedictions: the planet's thin mantle abrupted and fell, swallowing entire cities. In his dreams everything was destroyed. There was only the hot amber sand again, the primordial Santa Ana winds turning through everything. Sometimes he lay in bed for hours, feeling drained of energy, as if in his dreams of destruction things had actually collapsed inside of him. Later he would sit on his porch glancing through secondhand men's magazines he had purchased at Phil's Paperback Exchange on Ventura Boulevard, the perennial television playing from the neighboring bungalow into the otherwise abandoned courtyard like some unscathed bit of technology after the soft, impactless explosion of a neutron bomb. In the bungalow across from his, a woman with a brittle voice and pale, dry hands would occasionally prize open her kitchen's dusty venetian blinds and gaze at him for a while, her television mindlessly promoting cereals and deodorants to the courtyard's stunned and diarrheic cats.

'What are you doing?' she asked one day. He could glimpse only smoky dust swirling behind her cracked blinds.

'Nothing,' he said.

'Are you the man that drives the bus? Do you drive the yellow bus?'

He thought for a moment. 'I don't think so,' he said.

A chattering pneumatic hammer started up miles away. 'I just want you to know, then,' she said after a while. 'I just want you to know that we all think you're doing a very good job.'

*

'The mind has special powers all its own,' Victory told him, sitting with him during her breaks at the Golden Pancake Coffee Shop on Lankershim Boulevard and smoking Tareytons. 'I believe that very strongly. I don't think you can ever get to know me really well unless you understand I happen to be a very mind-oriented sort of person. A lot of people aren't mind-oriented sorts of persons at all. A lot of people just think money matters, or having a nice car.' She never really looked at him, but rather out the window at the flashing sun. A painting on the coffee shop window depicted a jolly fat chef staggering underneath a platter of enormous pancakes and syrup. 'That's why I came to LA, you see. Because LA is the land of the next great superrace. There's a conjunction of planetary currents and influences here which are very powerful. That's why Aimee Semple McPherson came here. That's why Krishnamurti came here.'

He drank his coffee and thought about enormous pterodactyls carrying people off in their bloody talons, withering droughts, vast clouds of hazardous waste settling over the city like a bird on its nest. And earthquakes. Sudden, impossible, perfectly enunciated earthquakes. They didn't speak to his mind, but to his bones. It was like the articles in supermarket tabloids about mothers psychically contacted by their unborn children. Unborn children wanted to tell you things. They wanted to prepare you for their imminent arrival.

Victory lit another Tareyton. 'You're a very good listener,' she told him after a while. 'Men are never usually very good listeners, you know.'

Victory's real name was Eleanor Davenport, and her father was a fundamentalist minister in Austin, Texas. 'Being raised by fundamentalists can be a very wearying experience,' Victory told him. 'There's no room for growth when you've got a big helmet on your head. Ever since I came to California I feel as if I have won the most important battle of my life, and nobody can ever tell me how to think or behave ever again. That's why I changed my name when I came here. To Victory.'

Victory had been living in Burbank for two years, sharing a two-bedroom block-style apartment with a legally blind Tarot reader who went by the name of Governor Pearl. 'She has taken the name Governor Pearl because it combines traditional masculine ideas with traditional feminine ideas, thus helping create entirely new patterns of socially accepted feminine behavior. A person's name is very important since it tells people who you really are inside. Otherwise you're just like everybody else. You're just a Mary, or a Jane, or a Bill. Basically I guess you could say Governor Pearl's a very thoughtful, caring sort of person who's hardly ever home, but pays half the rent anyway.' David never met Governor Pearl, but a few times Victory pointed her out to him as she drove past the diner in her long white '57 Thunderbird convertible, maneuvering slowly through the crowded street traffic, her guide dog, Jeff, panting beside her in the passenger seat. Governor Pearl wore dark sunglasses, and had a long gaunt face with close-cropped hair. 'She's not really completely blind,' Victory said. 'She can still make out general shapes and things.'

Usually they spent the night together at his place, lying naked and pale on the sweating mattress, gazing absently at the thin gauze curtains as if expecting, at any moment, some indication of a breeze. Victory loved to talk, and even when she didn't have anything to say she had numerous questions to ask.

'Do you ever think of spiritual things, David?'

'I don't know. I guess so.' He could still hear motorcycles echoing in the courtyard, tearing and shaking at things like tremblers beneath the world's floor.

'Like what sort of spiritual things, David? Do you ever wonder about your rightful place in the universe? Do you ever wonder why you were born, and what your purpose in life is? Do you believe there are spiritual beings who have come from millions of years in our future to determine our lives and thoughts, residing underneath our world because they fear our violence and narrow-minded hypocrisy? Or do you believe we all have free will, and that each of us is a God of his or her own private universe, the universe of the self? In our private universes we live forever. Nothing restrains us in the universe of our minds, because that is the real universe, and this one that we think is real is just a prison where we go when we want to punish ourselves. Do you ever think about things like that, David? Is that why you're so quiet all the time? Have you ever read Jung? Or Gurdjieff? Have you ever been to a phrenologist? Do you believe in phrenology?'

Sometimes David grew dizzy and disoriented by Victory's questions. They seemed to multiply in the tiny, hot bungalow like strange misshapen insects with chitinous bodies and jeweled glistening wings. They fluttered and banged against the naked overhead light bulb; they crawled into your clothes and sheets and facial hair. He developed a flickering migraine, and thought as hard as he could. 'I don't know,' he said finally. 'I don't know if I've thought about that stuff or not.'

Victory was convinced David should meet her Spiritual Work Group counselor, who owned a house in Sherman Oaks. The house was filled with paperback books on tidy mahogany shelves, energy crystals, pyramid tents and totems, and original Native American religious artifacts stored behind polished, glass-paneled cabinets. 'It's like a museum, kind of,' Victory said. 'It's like a museum of higher thought.' She pulled her battered Toyota Corolla up on to the slender curb. 'He's giving one of his famous High Energy Workshops today. People come from all over the world to attend them. But he's graciously agreed to meet us both afterward for a short while.'

'Intellectual abilities are not something to be taken lightly,' Dr Simonson said, drinking orange pekoe tea from a large ceramic mug. The mug was imprinted with a blazing red and amber mandala. 'With knowledge comes great responsibility, and I don't mean responsibility to mere mankind, for we share our universe with creatures of many different species. There are creatures on Rigel-seven, for example, who breathe pure methane. They have evolved many thousands of years beyond our puny race. Because they dispensed with violence long ago, they no longer possess hands or feet. They make love by means of pure meditation. Often lovers are together for thousands of years without ever meeting face-to-face, without ever knowing what each other looks like.' They were sitting in the large ornate living room which was crowded with more bookshelves, enormous fetal stone sculptures, and an intricate stereo system with thousands of records filed in varnished oak cabinets. The room felt infinitely heavy, David thought. It felt like massive tectonic plates jostling underneath continents, oceans, gravity. David thought he felt something shifting in all that buried weight, tearing. He placed his hands under his knees.

'Ideas belong to the entire universe, David. That's all I'm saying.' Meanwhile, David tried to envision the ruptured walls of Atlantis, the ocean pouring through and pounding everything to bits. Enormous cyclopean squid pulled away, pumping like hearts. Strange convoluted creatures with bloated symmetrical features. Vast green plains of algae and rocks, dark intricate caverns filled with unglimpsed and impossible life forms. Nobody could hear or see anything down there. You could only feel it, that sudden *beat* of the sea. The superrace, with their nightmarish faces and inflated, brainy skulls, had long anticipated everything, even their own destruction. The world of the mind goes on forever, they thought. And the world of the mind is all that matters.

David felt a sudden, weightless sense of relief rise in his chest as he considered the earth's doomed secrets. Then, after a moment, the plates in the cupboards began to rattle. The large bay windows chattered faintly in their frames like whirring insects. Victory clasped shut her purse.

'Don't worry,' Dr Simonson said. 'It's just a little trembler. We get them all the time.'

After the destruction of Atlantis the sea looked exactly the same. A few jostling whitecaps, waves. Gulls screeched and wheeled, clouds drifted overhead. On the ocean's surface it was as if nothing had happened.

'Earthquakes are very important psychic occurrences, you know,' Dr Simonson said. 'Earthquakes help the world breathe. They adjust the world's psychic energy so everything can flow smoothly again through our eternal minds.'

'This is the guest room,' Dr Simonson said. 'This is my office. That's an original Dali, a Braque, a Chagall. Great art gives us knowledge of the soul. This is my *Marvel Comics* collection. This is my science fiction room. My girl comes in four times a week and cleans everything, even the windows. Here's where I keep my pinball machines, video games, hot tub. All of these bay windows – I had them recently installed myself. I commissioned a semi-Olympic-size swimming pool for the yard. I'm going to tear down the garage and the guest bathroom and build a tower, which I will fill with my metaphysical library, my Indian and East African religious statues and a massive, a really humongous Dolby sound system. When I play Strauss, it'll sound like a football stadium. It'll sound like the entire place is lifting off into outer space.' By this time they were back in the living room and Dr Simonson was offering them ginseng tea, Chinese almond cookies and Hershey's Chocolate Kisses.

'When I came to LA in 1972, I had nothing,' Dr Simonson said. 'I was driving a '63 Chevy Nova. It had three bald tires and no brakes. If I wanted to stop, I had to downshift into low, then pull out the emergency brake. I was drinking two quarts of Albertson's whiskey a day. I was going nowhere fast. I was going to hell in a handcart. Then I met a girl on the beach at Venice and told her about my dream. I dreamed of an institute that wasn't an institute. I dreamed of a corporation that wasn't a corporation. I dreamed of a radical organization which wouldn't get into any trouble. I dreamed of a group of people sharing their noblest thoughts and ambitions. I dreamed of all this.' He gestured abstractly with his teacup. The bay windows, library, contracted swimming pool. The San Fernando Valley lay spread out in the distance, dully glittering like an enormous transistor component. 'Her father was Andrew McLanahan, senior vice president of the Fluor Corporation, and she had a trust fund she wanted to do some good with. She believed in my dream. As a result, my dream became a reality.' He poured more tea, gazing absently out the sparkling bay window. The entire valley was bleached with smog, as if everything in the world were fading into whiteness around them, everything except for this room. Everything except for the statues and books and records and shelves. Everything except for Dr Simonson and the Worldwide Institute of Higher Learning, Ltd.

Some nights he took long walks through Burbank, Van Nuys, Encino, even White Oak, trying to recall his life before LA. It was all very cold and indistinct, as if it weren't really his life at all but just some movie he had seen years ago on TV. He saw himself sitting in the backyards of strange houses. The houses were owned by men and women named Nancy and Bob, Dawn and Phil, April and James. Sometimes he sat in their cool beige living rooms and awaited their questions. 'Do you like it here in Lompoc? Do you miss your friends? Would you like bologna and cheese, or ham salad? Do you like your room? Do you have any hobbies? Do you ever like to read a good book?' Then the men would take him into the garage to see their power drills, shortwave radios, miniature trains. He liked to watch the trains go around and around on the hissing track, and sometimes the men said, 'You can stay and watch for a while, but don't touch anything. We'll call you when dinner's ready.' And then leave him in the dark garage to watch the trains run around. He was afraid he might touch something by accident. Then the entire plyboard frame, plastic trees and houses, tracks and cars and station houses would all fall crashing senselessly to the concrete floors. He knew he shouldn't touch anything. Maybe he shouldn't even look. Whenever things were broken or missing, they always asked him first. Trevor and Sally, Alex and Mary Ann. They were always looking at him as if they could see broken or stolen valuables reflected in his eyes. That's why he preferred to sit alone in backyards, where there wasn't anything valuable he could damage or destroy. If he sat very quietly, the women wouldn't ask him any questions, or make him any more sandwiches. If he sat very quietly, they might forget he was even there at all, as if he were the brick barbecue, the high tessellated birdbath, the thorny rosebushes bound to thin green bamboo poles. There were Indians he had read about who could sit so still you couldn't see them, even if they were right in front of you. He would hold his hands clasped in his lap, and open them slightly from time to time, careful not to let too much air out. When he clasped his hands together they formed a wind box, and you could hear the wind if you held your ear close. He liked to hear the wind and keep it safe there in his hands, where it wouldn't disturb anyone, or attract any attention. An old woman had taught him the trick in one of the first houses where he had lived. The woman had made liverwurst and cheese sandwiches. Tomato soup. Chex Party Mix. He couldn't remember her face anymore, but he could still remember the liverwurst and cheese sandwiches, tomato soup and Chex Party Mix.

Victory gave him Gibran's *The Prophet*. She gave him Tolkien's *The Fellowship of the Ring*. She gave him numerous paperback novels by Philip K. Dick. The books had broken spines and were thumb-soiled. Their lurid covers depicted robots, embattled spaceships, babies drifting through space. 'Philip K. Dick is the greatest American writer of our century,' Victory said. 'He was the only great writer to understand what a big and amazing place our universe is, as big and amazing as the darkest places of our own minds.' Twice each week she brought him to her Spiritual Work Group at Dr Simonson's house, where he was encouraged to share his latest 'ideas.'

'Always remember,' Dr Simonson urged them, solemnly igniting sandalwood incense and pouring the inexhaustible ginseng tea, 'all ideas are good. No ideas are bad. There are half-formed ideas and undeveloped ideas. There are ideas which you don't understand, or which don't make you comfortable. But remember: We are a collective human enterprise of this Planet Earth, competing in the universe for greater knowledge, more perfect comprehension and truer love. Even as we speak, our thoughts and ideas are being projected into the living rooms of Tau Ceti and the Omar Asteroid Belt. In Tibet, men of vast religious wisdom have ingested hallucinogenics in order to mentally transport themselves here to Sherman Oaks and thus monitor our secret progress through the world of intellect. Would you like to start, David? Since you're our new member, would you like to give us a little information about your intellectual life?'

He wanted to say yes. He wanted to tell them about the destruction of Atlantis and the imminent destruction of LA. He wanted to tell them about the wind box, the smooth, taut skin between Victory's shoulder blades, the white moon pulling every evening at the deep earth. But all he could say was 'I don't know. I don't think so.' He couldn't even look them in the eye when he said it. He looked instead at this hands folded in his lap. The wind in the box. The wind didn't move if you held it very carefully. 'I guess I can't really think of anything,' he said.

Some nights, at home in bed, Victory tried to draw him out.

'You have to learn to share, David. I don't mean this as an insult, but when you don't share your secrets, you're acting like a very selfish individual. What would the world be like, David, if nobody ever shared their feelings with anybody? It would be a pretty cold place, wouldn't it, David? Don't you think it would be a pretty cold world we lived in?'

'I guess so,' he said. He thought it might be like Antarctica, covered with ice. You could wear heavy clothes in Antarctica like a disguise. He thought he wouldn't mind living in Antarctica.

'Tell me, David. Tell me how you feel.'

'I don't know.'

'Of course you know, David. How do you feel? How do you feel about me?'

'I like you, I guess.'

'You *like* me? You *guess*?'

He thought for a moment, watching his Marlboro smolder in the glass tray. He picked it up. He shrugged. 'I guess I love you. Is that what you mean?'

One evening after work Victory brought home a Feel Wheel in a large cardboard box. It was packaged like a board game, and contained a cloth sheet partitioned like a dartboard. In each varicolored segment was stenciled a different human emotion. Love. Sadness. Hate. Sorrow. Rage. Embarrassment. In the middle of the design, where the bull's-eye would have been, it said, 'I need to be left alone.' Victory handed him a red plastic checker.

'Now put the red checker on the emotion you feel most,' she told him. 'Only you can't put it on the center one this time, because, see, I already put my blue one there.'

They smoked some marijuana. David felt cool and blurred. The deep earth seemed very close and compact.

'So why do you feel rage, David? Why do you feel so much rage all the time?'

'I don't know.'

'What does it feel like, David? How do you feel when you feel rage?'

'It's like steam in a kettle,' he said. 'It's like water in a garden hose.' As he began to talk he grew slightly disoriented. He felt as if he were fading away. As if the ground underneath his feet were slipping, slipping. 'Sometimes I even travel down there, you know. I move among plates of stone and basalt, past sunken lakes of oil and natural gas. The stones are etched with skeletons of prehistoric dinosaurs and men. Enormous fish with teeth and tiny atrophied arms.' As he talked, he tried to recall the silence which had resided in his bungalow before Victory. He felt as if that silence were being threatened by the pressure gathering along this deep earth's trembling faults and fissures. 'Trilobites like intricate snowflakes, enormous plugs of lava, brilliant veins of gold and silver, and then this terrific pressure just building. I can feel it in my neck, my back, my teeth. I get these sinus headaches. I'm afraid to move or touch anything, because I know I have to be careful, I have to be careful with other people's things. And then other times, you know, I *want* to reach out. I *want* to give it a little push.' A sudden weightless sensation filled his chest and sinuses like a gasp of helium. He felt slightly panicked, as if he had been discovered in some embarrassing act by strange people who would post him off to other, stranger houses. Then he heard the dishes rattle in the kitchenette. An aluminum pot clattered on to the floor.

'Did you feel that?' Victory asked. She was leaning back, her eyes dilated and stunned as if by one of her own insights.

'It's like a dead ocean down there,' David told her. 'It's like Atlantis. There are entire planets down there that have been buried for millions of years. Sometimes they just need a little push. Sometimes they just need you to notice they exist.'

The next night after one of his longest walks he could feel the muscles taut and alert in the backs of his legs. He stopped at 7-Eleven for orange juice and Kit Kats, and arrived home around ten p.m. to find Dr Simonson in the kitchen washing an Underdog coffee cup in the cracked sink. Victory was on the floor leaning back against enormous frayed pillows, reading a paperback copy of Jung's *Man and His Symbols*. The third person in the room was a man. He was sitting on the sofa. He was tall and gaunt, with large, thick-rimmed glasses. He held a large gunmetal black box in his lap. He didn't look up when David entered. He seemed to be thinking about something.

The tiny cap fitted neatly over David's head and was attached by coiled wires to the black box.

'Alpha waves record the mind's mental activity,' Dr Simonson told him. 'The alpha wave is the dominant strain in intellectual thought. William Burroughs knows about it. Timothy Leary knows about it. The world's greatest advances were not pioneered by scientitsts but by artists. Galileo, for example, was not an astronomer but a poet of abstract space. Don't be afraid, David. There's really absolutely nothing to be afraid of.'

The thin man adjusted dials on the gunmetal box. Then he attached another thin wire to David's wrist with a Velcro band.

'We are all poets of infinite space,' Dr Simonson said, holding his warm teacup between his hands. 'We all deserve worlds as beautiful as the worlds we live inside.'

Sometimes they sat for hours in David's apartment while the gunmetal box emitted its endless ribbons of tape. David felt very uninvolved with the entire process, smoking cigarettes and gazing aimlessly through the books and magazines Victory brought him while Dr Simonson discoursed about civilization.

'Every individual has a certain skill or talent,' he said, slicing Gruyère cheese and attaching the fragments to whole wheat crackers. 'Every individual has certain responsibilities to mankind as a whole. Some men are strong, or beautiful, or know how to fix an automobile.

Some are good at math, or sculpture, or computer programming. But some men have special skills which you don't come across so often. These men are the guiding forces of civilization. Everybody in the world is equally important, David, but these men I'm talking about are *really* important.'

Sometimes, gazing at the smoky television screen, David let his mind descend into the notched and convoluted maze of the deep earth. The dead planets were still there. The patient faces of special people and superadvanced races. They all kept very quiet now that Dr Simonson was listening. They, too, heard the whirring of Dr Simonson's machine, and Dr Simonson's voice conferring with the thin man.

'What were the readings like?'

'I can't tell.'

'Did you check them against the seismograph?'

'Well, there's correspondences, I guess. But we haven't had any strong variations. We need a couple of good jolts first.'

They were looking for something. Sometimes it seemed as if they were looking for something inside David.

'David?'

'Yes.'

'How do you feel?'

'Okay.'

'A few days ago, when you were with Victory. How did that happen? What did you do?'

'I don't know.' David had made one firm resolution. From now on he wasn't telling anybody anything.

'What did it feel like?'

'It felt like everything moved suddenly. Everything started shaking.'

Dr Simonson's eyes were very black, David thought. After a while Dr Simonson said, 'I mean in your mind, David. What sort of intellectual focus did you have? Did you hear voices? Did anyone tell you what to do? Was it anybody you recognized?'

David thought for a moment. 'I don't know,' he said. 'I don't think so.'

'It was very . . . I don't know,' Victory said. 'It was very spiritual. I guess that's all I can say. I could feel the energy emanating from him and I knew he was a very special sort of person. I knew David could destroy everything if he wanted to, but he's far too benevolent and caring a person to do anything like that without a good reason.'

At night it seemed as if nothing unusual had ever happened. There was just him and Victory on the warped mattress, listening to cars in the street, with no real memory of Dr Simonson other than the strange strategic shapes of mandalas and pyramids and Navajo totems distributed throughout the musty bungalow. When it was very late David would let himself quietly out of the house and resume his restless walks, feeling all the world's pressure gathering underneath his feet. One night as he was letting himself out he found Dr Simonson sitting on the bungalow porch in the cat-clawed plastic lawn chair. His white penny loafers were propped on the wobbly wooden porch railing, his hands crossed in his lap.

'I'm not really Dr Simonson, David,' Dr Simonson said, turning to look up at him, the glare of streetlamps reflecting off his glasses and transforming them into expressionless white disks. 'Actually, I'm just Dr Simonson's egoic projection. I'm the man Dr Simonson always dreamed of being. Calm, self-assured, loving, immensely intelligent.' He was wearing pleated white cotton slacks and a loose-fitting beige shirt. 'You just go ahead with what you're doing, David. When I'm in this condition, I just like to drift along. You pretend I'm not even here at all. Remember that I'm still asleep in my bed up in Sherman Oaks, okay?'

These were hard desperate days of terrific heat. Temperatures and humidity soared to record levels, and at night David's walks grew longer and more furious, as if he were trying to outdistance himself. Interminable streets with savage, impossible names like primal litanies. Lankershim, Tujunga, Sepulveda, Vanowen. He walked for miles and miles, often conspicuously trailed by Dr Simonson, who, blocks away, strolled along imperturbably like a white cloud. Sometimes Dr Simonson drove his immaculately buffed Eldorado in the street beside David, playing Strauss's *Der Rosenkavalier* on his compact disc player, his windows rolled down and his convertible top popped back. Sometimes Dr Simonson turned down his music and talked with David.

'You probably don't believe this, David, but I have felt great anger and resentment in my life too. I have felt tremendously violent hatred against the world, have desired terrible retributions for the many cruel crimes committed against me. I know I should be more understanding, but sometimes I'm not; sometimes I'm not very understanding at all. Just a week ago, for instance, the Internal Revenue Service froze my corporate accounts. They subpoenaed many of my most loyal students, my secretary, my entire accounting firm. They're trying to declare me a profit-making institution, rather than a nonprofit therapeutic health maintenance community service, which is, of course, what I really am. Sometimes I'd like to let them all have it. Wouldn't you, David? Wouldn't you like to teach them all a lesson they'd never forget? All those goddamn bureaucratic phonies downtown. They're all asking for it, aren't they, David? They're all asking for one killer jolt, and you're the one that could do it. Let them die. Let them all be crushed and die. Then they'd know how it felt, wouldn't they? Then they'd know how it felt to build something you loved and have it all torn down around you by greedy vultures.'

The pressure and heat were growing intolerable. David suffered painful sinus congestion, recurring flus, chills and colds, sudden flashing headaches unrelieved by aspirin or Tylenol. Tension accumulated in the earth's secret faults and caverns, the deep earth filling up with resentment, like some cheated lover. It was all perfectly natural, he thought. In earthquakes natural terrestrial rhythms emerge from the earth's heart. Crippled women were known to walk after earthquakes. Blind men saw. Diseases went into spontaneous remission. Some great eternal pulse was always restored by cataclysm. 'You could let it happen,' Dr Simonson said. His eyes were damp and watery with allergy, his nose red and chafed by his pocket handkerchief. 'You could give the world back its heart, its breath, its voice. It's like cleaning out all the world's hate with vast white rivers of pure energy.'

Then one night he destroyed Los Angeles over and over again in his dreams. Deep subcrustal plains slipped and heaved, enormous fractural fissures opened across the entire San Fernando Valley, swallowing streets and housing tracts, ripping out interminable colonic streams of plumbing and electrical wires. Gas and water mains burst, filling the air with glistening sprays and roiling white clouds. Subterranean toxic waste containers burst and sprayed. The San Andreas Fault gave a sudden pull, and then, with a long, slow and almost graceful flourish, the great fragile promontory of Los Angeles snapped and slid into the boiling Pacific like a string of pearls down the grate of a sink.

David awoke feeling a strange lapse in his stomach and then, faintly, what might have been the subsiding aftershocks of an earthquake. He looked around the room, but everything seemed in place. The bungalow was quiet, and the streets outside quiet too. It was almost eight o'clock. Victory had left for work, and the bathroom was still misty from her shower. David lay half-awake in bed for a while. Something seemed very wrong. He picked up the telephone and heard a dial tone. He turned on the television and saw *AM Los Angeles*. After a while he pulled on his light flannel robe and went out to the porch with a cup of instant coffee.

Outside he couldn't tell if the world had ended or not. Perhaps *AM Los Angeles* was on tape delay in some collapsed Hollywood studio while its actual stars and guests struggled hopelessly underneath fallen girders and masonry. The entire courtyard was littered with wrappers and Styrofoam fast-food containers blown in by the night's muggy Santa Ana. The front end of the crippled Buick had been dismantled, baring rusted axles and joints, corroded drums and cracked brake shoes. The horizon was cloudy with either smog or the dust of pulverized buildings. He felt deeply exhausted, as if he had been running great distances in his sleep. Gradually he realized that someone was watching him. A large grasshopper whirred and chattered and landed by his feet. He looked up at the blinded window of the neighboring bungalow.

'Did you feel it?' she asked. 'I was awake all night. I was waiting for it, you see.'

The grasshopper looked at him with glassy eyes.

'It's God's fist coming down,' the woman told him. 'That's where earthquakes come from. From God's fist punishing sinners.'

David found a neglected cigarette and book of matches in his robe's vest pocket. The cigarette was frayed and slightly cracked along the seam. He lit it anyway. The grasshopper took off again and crashed into a nest of discarded Wonder bread wrappers. After a while, the neighbor's television started up, and someone on a morning talk show called it a three point seven, and someone else on the panel said, 'A three point seven? That didn't *feel* like a three point seven, did it? It felt at *least* like a four point *five* . . .'

*

David was alone in the house all day, smoking a pack of Marlboros he bought at 7-Eleven, watching daytime movies and reruns of old sitcoms.

'They've arrested Dr Simonson,' Victory told him when she got home, breathless, her uniform stained with yellow egg matter and ketchup. 'Tax fraud, David. I have to go up to his house before they've sealed it off and get his papers for him. I can't believe this. I can't believe they're doing all this to a nice man like Dr Simonson.'

The rest of the summer was much like the beginning, leaving David the bungalow to himself. Victory was occupied with work and grand jury proceedings, and then, in early September, Dr Simonson was admitted to the Betty Ford Drug and Alcohol Rehabilitation Center in Palm Springs, where Victory regularly drove herself and other workshop members to visit him. David found a job pumping gas and tuning automobiles at a local Gulf station, and in the evenings after he returned home he would sit on his porch and breathe the gas fumes from his soiled green uniform. The thin man with the glasses never returned, but the gunmetal black box remained on the plastic coffee table.

A cool sea breeze had begun developing by the time Victory left for Iowa. 'Father mailed me a ticket a few weeks ago, and I didn't know if I wanted to go or not until now. I've been thinking a lot about God and family lately, David. I've been thinking maybe it's all right to rebel when you're young and confused, but that later, when you get older, it's time to settle down and raise children so you can teach them about God's love. If Dr Simonson or anybody from the workshop calls, tell them you don't know where I went, okay? I hope you're very happy and find God in your heart too, someday, David. I hope what I'm doing doesn't hurt you.'

It didn't hurt him. The next few weeks were filled with a high overcast, muggy and clinging air, a certain heaviness and density of limb and brain. There were no more discernible tremors, but only a sort of faint and misty anxiety which infiltrated news features, community council hearings and all-talk radio. Most of central Los Angeles was inadequately constructed. People recalled the St Francis Dam in the Santa Clara Valley, built with graft and faulty cement by William Mulholland. Los Angeles was just waiting and vulnerable. It would collapse like matchsticks. The Los Angeles 'natives,' however, refused to be intimidated.

'All-Talk Radio. You're on the air.'

'Hello, Linda. This is Danielle in Whittier. I just wanted to say, about all this fright talk about earthquakes? Well, if you live in the Midwest you've got your tornados. And if you live on the East Coast you've got your flash floods and hurricanes. There are electrical storms in Kansas my boyfriend Gary told me about? He saw this enormous ball of electricity rolling down the freeway and it hit another car and the car exploded? Well, I'm just saying, earthquakes aren't so bad, after all. I mean, they're here and everything, so what can we do, right? Oh, and Linda?'

'Yes, Danielle?'

'Do you really think old Teflon pans could be dangerous for frying foods? My sister Betty gave me her old set, and it *looks* perfectly fine and all. I'm just worried about the kids.'

He never saw Victory again, or Dr Simonson, who was released from the Betty Ford clinic in October and, after an elaborate court trial which lasted nearly two years, acquitted of conspiracy and fraud. After paying a substantial fee to the IRS he reportedly moved to the Bay Area where he organized the expansion of his Free Spriritual Outgrowth Clinics into Arizona, Hawaii and, eventually, even South Korea. Sometimes, in his walks through the Valley, David noticed clinics springing up in disused shopping and medical-dental centers. Lovely blond girls in overalls gave him pamphlets as he passed. He never really looked at these women, though. He preferred to look at the beautiful women driving by in their cool, expressionless convertibles.

Three summers later he was hitchhiking home from work on Ventura Boulevard when Governor Pearl pulled up and offered him a ride in her white Thunderbird. 'I enjoy driving at night,' she said, turning to look over both shoulders. Without sunglasses, her right eye appeared lifeless and dim, her left eye overlarge and milky with glaucoma. As she pulled slowly away from the curb a passing car honked and veered. In the backseat her guide dog lay unharnessed and senselessly asleep.

'I love to drive in LA,' Governor Pearl said, and offered him a Coors from a white Styrofoam icebox, Winstons, the electric cigarette lighter. Her right eye, trained on the road, tilted her head to one side. 'There's probably some chips or something in there too,' she said, gesturing at the large brown 7-Eleven bag cluttered with wrappers, unopened toilet paper and a discarded Big Gulp container filled with dissolving Orange Slush.

They took the Ventura Freeway to Santa Monica, then the winding and exorbitant Sunset Boulevard back into Hollywood, Silver Lake, Chinatown. 'It's so big,' Governor Pearl said. 'And filled with so many interesting people. Is that a stop sign ahead?' They slowed to a stop. It was a DON'T DRIVE DRUNK billboard. They got back on the freeway and drove until dawn, drinking Coors and smoking marijuana. Around seven a.m. they stopped for Egg McMuffins before dropping David off at his home.

After that night, David completely forgot about the deep earth, Dr Simonson, Victory, the wind box, his many concerned foster mothers and foster fathers. It felt like going away. It felt like awakening to a new life in a new house with strange people in it and just going away from all of them, because it was never a place you belonged anyway. The world was very peaceful after you went away. Everything was new and fresh and clean. Everybody breathed easier, thought more clearly and lived longer, healthier lives.

In the warm regenerate spring he abandoned everything and moved to the Pacific Northwest, where it rained for him nearly every day.

CLOSER TO YOU

In the sunny mornings you chase the orange cat across the green grass. Underneath your feet the shelter, filled with the moon, the rain and the stars and the wind. And the orange cat lands on the fence, its orange tail twisting. Its dead sparrow on the crumbling brick barbecue. Big weeds from the barbeque, and the rusting iron gate, and brown hollow snails all around. The dead sparrow says when you touch it. The yard says, the underground shelter says   The underground shelter filled with everything that is not you. And the dead sparrow says to your fingertips, your tongue and your mouth, and then mother on the back porch shouting the No! The No stella! And you are no longer stella, and the dead sparrow falls to the green grass, its words still feathery and wet on your thick tongue. The No! in the air and the orange cat on the red fence, its orange tail twisting. And you think the No! at the cat, and with your mouth you say   With your mouth you say   The cat's orange tail twisting. And the dead sparrow on the green grass.

Breakfast time and you are stella again, every morning surrounded by the world that is not stella. House, sun, moon, table, cocoa puffs, liquefying margarine, sprays of black bitter crumbs, mother, marcie, tablecloth and dad. The grandma on the stuffed chair, with her food tray and cracked plates. And mother holds the big spoon. The grandma says Aaaah. Aaaah, with gerber peaches on her chin, and mother says Here you go, alice. Be a good girl, and dad says There she goes. Now she's started. Dad doesn't like breakfast and sits on the sofa, wearing his blue robe with the torn blue pockets. And says I thought I detected a minute there. I thought I detected an actual minute of peace and quiet in this goddamn house, and the grandma says Aaaah. Aaaah. So mother gets the medicine and you drink your juice. Orange juice in mother's glass. And the marcie clacks her new pink shoes together, requiring more toast. The marcie wears red paint on her face today, just like mother, and the voice on the radio says Better. Brighter. More Beautiful. The noise of the world on your face. You can touch it with your sticky hands. The noise of the world that is not you, and then you say   And then you say

Dad says Great. Now they're both started. It's like a chorale – one vast great vegetable symphony, and takes his coors out to the sunny back porch.

The grandma's tiny eyes are red and wet, filled with words and memories of words. She says Aaaah. Aaaah, her big veiny tongue extruding from her thin lipless mouth. Her mouth red and glistening, but not like the marcie's mouth, not like mother's.

The gerber peaches are all gone, and now the grandma loves you even more. You and your soggy toast.

And so you give her some.

Every weekday morning mother takes you to the special school for special children. We just hope and hope, mother says. We just hope for the lord's compassion and his love. You hold the seat belt's bright silver buckle in your hands. The silver buckle is a big hard word, and pulls the belt very tight. Mother says the lord words again and again, and you put your hands together to speak the lord words too, feeling them between your fingers. Then you get out of the car at your school filled with special children. Big yellow bumblebees on a big wooden sign, smooth and you like to touch. This is your bright smile, this is your round stomach. Tea, tie, toe, tum, tah, tea mrs evans says. Her bright red fingertips touch your teeth, your tongue, putting the words there. Tih. *Tih*. With the *t*ip of your *t*eeth, with the *t*ip of your *t*ongue. So you tell her the grandma word and she gives you a green m&m. And she says now with the *tih* sound. Taaah. Taaah. Tooth, tongue, tea, taod, tot, and shows you big pictures in her lap. A big green toad, a cup of hot tea. And you say the grandma word again, and mrs evans has more m&ms. Red, yellow, black, orange, blue, brown. Because the grandma is a tongue, a tooth, a toad. And all the other special children make grandma sounds too in their big bicycle chairs with their broken eyes and big shiny foreheads, and you're all tea tie toe tum tah tea, and scrub the stiff colored paper with bright crayons. The crayons snap when you break them, just like mouths. You play in the tanbark yard, bang blocks to music, eat graham crackers, and dad picks you up in his car with a coors in his lap. Who's my little dollface, he says. Who's my little cracked brain, my little wordless wonder. Kissing you with his rough face, taking you home to drink more coors on the slanting-sunny back porch. This is your yard. This is your bright sun. This is your green, green grass.

Dad built the underground shelter in the yard because he loves you. In wars words fall and explode. In wars everybody falls wordless to the ground, wordless just like you. Usually wars are on television, but when they become real, dads dig holes in the ground. Big machines come into your yard and eat it. Strange men with hairy arms and beautiful tattoos. But then the war doesn't come and only you are stella, only you and the wordless people on TV. You like to stand on the hard wooden lid. You like to point at the big steel pipe filled with spiderwebs and shrouded, mummified bugs. The hard wooden lid fastened with a big rusty padlock. And then you ask dad And then you ask him And dad says I built the goddamn thing so when the thermonuclear war comes and blows us all to hell and back we'll have a warm place to go to the bathroom. Now give dad a kiss and leave him alone. And dad's rough face says kiss. Kiss kiss, it says. How'd I know they'd fucking lay me off? How'd I know I'd blow all our savings on that goddamn water trap and the bastards would lay me off? And pulls his blue robe tighter and cinches his blue flannel belt, just like the seat belt in his rattling car. The same car they drove the grandma home in when the grandma lived in a hospital with the world's other grandmas.

The marcie has her own bed but at bedtime she lies on your bed to play. The marcie plays like mrs evans and points, reading you the big red book and says This is the spider, the stone, the snake, the sock. And her finger says Point point point. Tomcat, teddy bear, toilet, towel, tripod, toad, and you are watching the goldfish in the round fishbowl. The goldfish mouth moves too, just like the marcie mouth, and you reach for it. The goldfish says   The goldfish says   And you say too

And the marcie says the No! and grabs your hand. *This* is a tripod, *this* is a toad. Now which is the tripod? Which is the toad? And you make your finger say Point point point, and the marcie takes your finger from her mouth and says the No! You want to touch the marcie mouth, the hard round words. You want to take the words in your hand and put them in your pocket with the masticated plastic soldier, the stray pink barbie slipper, the broken crayon. But the marcie says the No! No stella! You're a *bad* girl and *nobody* likes you. Now go to bed now! Now go to bed now!

At night the grandma sleeps in the blue room, strapped to the crooked bed with the car straps, her eyes wide open sometimes. They are very hard clear eyes, eyes eyes, and the moonlight through the curtains talks in her room, but the grandma says nothing. And you say nothing back, wearing your soft pink pajamas. The marcie asleep in your bedroom, mother and dad asleep in theirs. You touch the grandma and she opens her big black mouth. Her thin hand touches you, grabbing. And you listen to the words in the grandma's head, like the words underground in dad's shelter. Spiders and glistening webs across the grandma's window, tiny bugs blurring and buzzing. The grandma sees the bugs too, the moon and the night and you. And the grandma's veiny hand grabs your pajama top tight and you can't get away. And then you whisper   And then you whisper   And when the grandma's eyes close again, you return to your bed and dream of dad's shelter underground, where words echo and resound, hop and crawl. If only you could reach them.

Some days you don't go to school and mother takes you to the doctor and his gleaming silver tools on a clean white cloth. And you say the grandma words and the doctor puts a mirror in your mouth. And mother says She tried to point yesterday. She pointed at the yard and tried to say something. An ess word I'm certain. When einstein was five he couldn't even tie his shoes, and little stella can tie her shoes. Mother shows you your shoelace and pulls, and so you fix it. And the doctor says Cognitive development, aphasia, phonetic, lexical and syntactic, langue and parole. And then mother takes you home to the yard where dad sleeps on the unsprung porch sofa, and she goes away again.

Underground in the shelter the moonlight's waiting and you walk across the hollow yard and through the hot sun. If you carry the old wood chair you can stand and peer into the rusty metal pipe. You can smell the buried life, rich with words and water and rust. You reach into the pipe with your hand and pull out ropes of dust, collapsed spiders and glistening white insect eggs. Those are the words down there, you can hear them in the pipe. Stone sea sand sister snake sun song soon. And one spider blossoms in your hand and starts to tickle. You hold it in your hand and it scrambles around, this thin tickling word. Ssssss. Ssssss. Some words go up and down, other words go back and forth. The spider is a word. The surf the sand the sea. And you push it through the pipe into grandma's world, where grandma waits in the dark. Spider. Spider spider. That's what you and grandma say together, underground where everything is wet and thick and real. Spider. And then you hear a round word, and under the roses you see it. It is big and fat and green. Ur, it says. Ur. It's wet in your hands when you lift it to the steel pipe. Toad, it says. Toad. A big green toad, a cup of hot tea. Ur. Ur. Falling down the long steel pipe where the other grandma waits for it.

Sometimes the grandma cries at night when you visit, strapped in her small bed, dreaming of the hard wordless night and all the world that is not stella. The grandma dreams with her mouth open, her mouth black and hollow like the steel pipe, her cheeks wet when you kiss them. Like the shelter, the grandma is filled with the world that is not grandma. Words like crumbling concrete blocks and broken red bricks, weeds sprouting and the unsprung sofa and the shelter's hard wooden lid. The grandma has an inside, too, and you touch the *t*ip of the grandma's *t*eeth, the *t*ip of her *t*ongue, and the grandma says Aaaah – the everything word. Aaaah. The moonlight burying itself tonight in the shelter, spilling down the thick steel pipe with the swift-gliding bugs. You reach into the grandma's mouth looking for words there, their hard glittering edges, like plastic toys lost in the overgrown yard. In the underground shelter, swamps and sculpted faces, pockets of fossilized bones and fuel, stones etched with prehistoric brains and skulls. Aaaah, the grandma says, her hands grabbing at you. Aaaah.

This is the secret here, she says. The secret sound of words, the secret dream of everything that is not stella. You love the word already as you pull it from the grandma's mouth, the grandma's eyes round and wide and glassy with something which is not moonlight. This is hard. This is real. You place the word on the *t*ip of your *t*ongue, salty on your fingertips. Your throat hurts the word, growling like a dog. 'Tor, taw, ter,' you say. And the moonlight pulsing in the shelter, and the toad hopping in the dark. 'Tord, tawd, toad,' you say, and that's it there. That's the word in your mouth now. 'Toad.' Its salty taste, its scaly skin. 'Toad.' And the grandma doesn't say anything, just looking at you. Very tired now, very sleepy. 'Toad toad toad,' you say. And the grandma says Aaaah very softly, very tired. Goes to sleep and her mouth falls open. Then you hear the other word there. Sssss. Sssss. Sea sand spider. Spider. Spider.

You say 'Toad' at breakfast and mother gives you french toast with syrup, and you eat the syrup with your tiny spoon, and jam with your toast. 'Toad toad toad,' and the grandma in her chair with her soggy food, dreaming of other words beneath the yard. 'Toad toad, toad toad toad.' And mother gives you a big hug, and the marcie gives you a hate face. But dad says It's original sin. My pure little brain case has fallen into the world of already fallen language. Great. More talk, more words. Everybody in the world will be talking someday. Today I think I'll look for a job. I gotta get out of this fucking house.

And mrs evans says Cognitive aphasia, positive reinforcement, syntactic redevelopment, and makes mother watch as she feeds you more m&ms. 'Toad toad toad.' And mother says I'm so grateful to the lord, and you open the big picture book and Point point point and everything is 'Toad toad toad.' And then the other thing in the shelter, the other thing tickling in the grandma's mouth. 'Spire,' you say. 'Spiner. Spider. Toad toad toad.' And everybody in the entire world loves you, just like mother.

There's no big rush, little cracked brain, dad says, you in his lap and the television on. There's no real hurry. *That's* all you're hurrying toward. *That's* all language is about. And the television says hostages in lebanon, preschool drug addiction, dioxides, acid rain and nuclear waste.

'Fish,' you say. 'Fish fish fish. Goldfish. Goldfish daddy.' You can still feel the wet squirming word in your hand. You can still see it falling down the long black pipe. And the marcie saying you're in trouble now, boy. You're in trouble now. The round bowl-water empty and opaque with tiny white feces. And next day mother says Where's our family photographs? They were right there on the mantelpiece. And the marcie puts her fists firmly on her hips to tell them, but then you say 'Mother. Daddy. Marcie. Grandma. Grandma.' All those words you found in the grandma's mouth last night. And nobody hears the marcie say a thing. Now it's the marcie's words that don't really matter.

'Kodak,' you tell them finally, just to let them know you understand even big words, too. 'Kodacolor. Kodachrome.'

But the grandma is never happy when you visit, looking at you with her big black mouth open. Aaaah. All the world's loud words resounding and spinning down there in the grandma's mouth. You try to pet and soothe her. These are the words, you think. These are words right here, and the grandma kicks and grabs, gurgling where your hand is, and so you take your hand back. Grandma grandma. All the darkness inside the grandma grows and grows, whispers and whispers, moves and pushes. Everything's better down there. Down there the grandma can be happy again. Down there the grandma can be grandma again.

You lead her down the hall steps. Doo, doo, doo, doo. One two two four two one two two. The grandma holds the banister because she's afraid. She makes different noises now, wordless noises down in her stomach and thighs and feet. Doo doo doo doo doo. All the shadows hanging from the walls and furniture and curtains, and opening the squeaky picture window, the grandma leaning against you, her body thin and frail and very soft like a giant stuffed giraffe. The grandma all hollow spongy bone. The grandma all sound and word and dream. Outside, the night is filled with stars and the big fat leaning moon, humming there, filling the steel pipe and underground shelter and grandma's black mouth. Aaaah. Aaaah. It is cold, the grandma wants to say. It is hot. It is cold and hot, big and small, boy and girl, happy and sad. Aaaah. Aaaah. Everything, the grandma says. Everything, everywhere.

The grandma doesn't like the steel pipe, the buried shelter, the heavy moonlight down there just waiting for her. You try to help her into the steel pipe. Down there, the other grandma reaches too. Down there the other grandma pulls. But this grandma says Aaaah. Scared but she wants to know. Always the deep earth, always the black night. The grandma is like the orange cat, which clawed and ran away. Grandma grandma. You try to make the soothing sounds mom makes for breakfast. There there Alice. Be a good girl. You like your cereal don't you Alice. Aaaah. Aaaah. Into the steel pipe, but both of grandma's arms don't fit. Pull her down, the other grandma says. Pull her down to me. And grandma's big black mouth wide, wide with all the words she'll find down there. Everything words. Aaaah. Aaaah. And the moon all around, and the words down there waiting with the other grandma. Louder and louder the grandma gets. Louder and louder like the moon. Aaaah. And now you try to help her. Help her head into the big steel pipe, but her head doesn't fit. Louder and louder, deeper and deeper, worlds and words. And lights going on in all the backyards, and lights going on in your house, too. Light, light. And the grandma crying for the other grandma, for all the words she can't quite reach when the men come with big steel tools and cut the pipe away, and then pull grandma from it.

Down there. Down there. Toads hop, snakes slither, spiders scramble and crawl. Moonlight talks and dark earth listens. Pih, pih. Paper, pin, pickle, paste, plate, pastry, pigeon, plant. Person. Person. And now the *r* words. Rih. Rih. Everything down there in the dark and the water. And grandma embracing everything too. Vast sunken cities, countries, landscapes and stars. Hissing rivers and steaming forests. Worlds down there, worlds and words. Rih, rih. Rock, robin, rouge, rasp, rattle. Dad's graduation ring. The marcie's transistor radio. And grandma, grandma coming closer. The grandma's arms big and strong and beautiful now, reaching to hold you, reaching to take you home . . .

You wake up.

You sit up in bed. Everything is black.

The marcie kicks in her bed, snorts and rolls over.

Outside moon talks and deep earth listens. Down there. Down there.

In the grandma's room across the hall, the grandma's bed is empty.

Grandma down there.

Grandma caught a cold, mother says at breakfast. Grandma caught a cold and went away.

Then after breakfast you all get in the car together and drive to the cemetery. Round brown slopes and green hedges and tidy, solid tombstones and flowers in little vases. This is grandma's new home, this big green yard. You all stand together under the low cloth awning and watch them lower the grandma into her new, dark room.

'A bad cold,' you say, and mother squeezes your hand. 'Grandma has a bad cold.' And everybody back into dad's car.

Mother snuffles into her handkerchief, and the marcie reads a *Doctor Strange* comic book. The King-size Summer Annual.

'Grandma goes away,' you say. And so you all drive home, where the other grandma waits in the darkness.

Maybe she's found what we all hope to find someday, dad says, and now you are all driving into the driveway of your big warm house. The dark clouds breaking apart. Bright blue sky blazing through. And you want to tell dad the secret as he lifts you under one arm and carries you to the front door. Come on, my little sack of potatoes, dad says. Come on, my little fat bag of laundry.

Dad is bigger, but you know the secret.

'Everything,' you whisper. 'Everything, Daddy.'

THE LAST MAN THAT TIME

'I'm going to tell you something, Mary – and I'm going to tell it to you straight. Because I respect you, and I think you're learning to respect me a little bit too.' Bernard Chalmers lifted his glass of red wine as the waitress set down his prime rib and Mary's chef salad. 'To be perfectly frank, Mary, I think you're one really hot-looking babe, and I'm not ashamed to say I've been having a lot of highly sexual thoughts about you ever since we first met. In fact, I've already lost track of how many sexual thoughts I've had about you in just the last hour or so, or ever since we met for drinks here at Pablo's Patio.' Bernard was watching the waitress's tan unstockinged legs as she carried her tray back to the kitchen.

'My,' Bernard said, never taking his eyes off the waitress's legs. 'This food looks really good, doesn't it?'

Mary Sinclair began crushing crackers into her salad. She was watching Bernard as if across a vast, lunar distance. His high receding hairline, beginning to break out in a pimply sweat. The thick protruding hairs of his nostrils, his flat yellow teeth, the dark splotch of red wine on his striped polyester tie. Mary was thinking, Definitely. The last one. I'll put everything into this I've got and then that's that. I'm not going out with another man in my entire life.

'I mean, we're both responsible adults, right?' Bernard was pumping steak sauce over his plate from a compact brown bottle as if he were dousing a small fire. 'I don't see why we can't relate to one another openly as highly intelligent individuals who also enjoy the unremitting heat of animal passion. It's perfectly normal. It's not like we have anything to hide.' Bernard was crosscutting his prime rib with a fork and serrated stainless-steel steak knife. Then he put his utensils down and gazed across the table at Mary. Uneasily, Mary returned his formidable eye-contact.

'So I've got this idea,' Bernard said. 'And my idea is this. Why don't you describe to me, in explicit detail, your last sexual experience with a man, and then I'll describe my last sexual experience with a woman for you? How does that sound, Mary? Do you think that'll break the ice or what?'

Mary was looking at the obloidal pieces of pink meat on Bernard's plate.

'Maybe I could finish my salad first – all right, Bernard?' Mary unfolded her heavy cloth napkin. 'Let me eat my salad first, and then I'll think about it.'

'So what did he say *then*?' Susan asked. Susan was painting her nails with flashing green nail polish. She was sitting across from Mary at Mary's butcher-block kitchen table. Mary had just poured herself a cup of steaming black instant coffee.

'So he said if I was feeling a little embarrassed, he'd go first and show me how easy it was,' Mary said. 'Then he started telling me about this girl named Christi he met at the laundromat. He asked if he could borrow some of her anti-cling sheets, and the minute they made eye-contact Bernard knew that their pheromones were really meshing. She took him to her house, threw their combined laundry all over the living-room floor, and had him right there, even though her roommate was in the adjoining bedroom with *her* boyfriend the whole time, listening to everything that happened. I mean, I couldn't really believe it, you know? You've got to *see* Bernard. Not that he's *bad* looking or anything. He's just not the sort of guy you want to rape on your living-room floor.'

'I guess,' Susan said. She was holding the nails of her right hand up for inspection. A person could flag down commercial airliners with nails like that, Mary thought.

'You'd be surprised, though,' Susan said distantly, walking among bright green fields and meadows, gazing across metallic-green rivers and salty green oceans. 'The sort of men women want to rape on their living-room floors, I mean. And the sort of men women never want to rape at all.'

Mary's problem was that she didn't know – she didn't know the sort of men she wanted to rape or not. When she got right down to it, she found most men – even quite attractive, muscular men with sandy blond hair and moustaches and stocky, well-developed leg muscles – generally disappointing at the animal level. Sure, she could enjoy nice meals with them, and even a few laughs. She loved dancing, and going to the movies and stadium sporting events. But whenever she and her date went home together it was impossible for her to take any of the sex part seriously. The men always seemed to be in such a hurry – and at other times they didn't seem to be in a very big hurry at all. They never picked up any of Mary's signals – and even when they did, Mary was never really certain what sort of signals she was trying to send in the first place.

'No, Bernard. It's not that I find you unattractive, or even undesirable. And, quite frankly, I've always considered receding hairlines a handsome trait in middle-aged men.' Mary had just finished her third glass of wine, and they were sitting together in large overstuffed leather lounge chairs in the bar of Casa de Julio in Tarzana. Mary was trying to concentrate on the empty, lipstick-smudged glass in her hands, because whenever she looked at Bernard he was either crunching more tortilla chips, or picking his teeth with the bar menu. 'It's probably just that I need a little time, that's all. I mean, it takes me a while to get used to somebody before . . . you know. Before I feel comfortable with them or anything. I guess you probably think I sound really stuck-up.'

'Like how *much* time?' Bernard asked, rummaging in the wooden bowl for more tortilla chip fragments. He had a distant, ruminative look on his face, as if he were trying to draw the winning lottery ticket from a gigantic steel drum. 'I know you can't be exact or anything, Mary. But, like, can you give me a rough estimate? Are we talking a week or *two* weeks? Or are we talking months, babe? Are we, like, talking years?'

Mary emitted a long sigh, and gazed off at an attractive, well-groomed young couple in the corner who were drinking jumbo margaritas and failing to keep their hands off each other. There was something cloying and insincere about them which made Mary angry.

She felt like getting up, striding straight across the bar and giving them both a good piece of her mind. 'It's people like you that make it difficult for the rest of us.' Mary wanted to tell them. 'So why don't you both just go home, do what you have to do, and stay the hell out of everybody else's face?'

Mary felt a silent, angular pressure filling the spaces between her and Bernard. When she looked up, he was gazing raptly at her with an unfocused, anxious expression, his moustache festooned with tiny yellow bits of tortilla chip.

Finally Mary felt the pressure about to burst. She wanted to get out of here. She wanted Bernard to take her home. No walk around the block, no nightcap, no nothing.

'To be perfectly honest, Bernard, I don't know how long. Maybe just a few days. Maybe a few years. Who the hell knows?'

Whenever Mary returned home from work at night she found messages from Bernard on her answering machine, or salacious gag-cards he had sent through the mail. The fronts of the gag-cards always presented a pink, droop-nosed, nude little man loaded down with beribboned gifts, candies and bouquets while he uttered sweet, practiced sentiments. Then, when you opened the card, large erect cardboard penises popped out, or condoms, or springy female breasts. Sometimes a wafer-thin transistorized organ beeped shiny, monochromatic melodies, and bold-faced black captions exclaimed, 'Ta-ra-ra-*boom*-de-yay!' or 'Dooh-day, baby! And I do mean – *DOO-DAH*!'

On the answering machine, Bernard's voice was always bright, sardonic and cheerful, as if he were being hurried back to some poolside press conference in Bel Air.

'Hey, sweetie. Guess who? Why don't you give me a call later, you know? We'll do something. *Ciao*.'

It was as if Bernard was filling her life with cuneiform, hieroglyphs, cyphers, the fragments of some untranslatable dream-life. Even after Mary burned his cards in the fireplace and erased the answering machine tape, something thin and irritant about the messages lingered in her apartment, distorting the relationships between things, abrading textures and smells like dust or bad karma.

*

In Mary's dreams, on the other hand, Bernard was a much pleasanter sort of individual altogether, with whiter teeth and a lot more hair. 'I guess I act like a jerk all the time because, well, I'm really insecure, and I suffer from an inordinate fear of rejection on account of my mother, being as she abandoned me when I was only thirteen.' Bernard was wearing clean white cotton shirts in Mary's dreams, tight faded Levi's and cowboy boots, and even resembled Sam Shepard a little now that he had improved his poor grooming habits. Despite his awkward exterior, Mary quickly realized that deep inside Bernard was intelligent, sensitive and thoughtful; all he needed was a little more confidence in himself. Then, just as Mary felt herself beginning to physically desire Bernard, she woke up.

'I guess the quality I admire most about myself is my sense of humor,' the undreamed Bernard said, sitting on Mary's loveseat and nursing the last drops of black coffee in his cup. Bernard had stopped by uninvited just as Mary was finishing her dinner and now he wouldn't leave. 'Too many people nowadays take the world too seriously. They're always worrying about their jobs or their families, or about some stupid war in Beirut, or some impending ecological disaster. And everybody's running around very sin*cere*ly trying to save whales and wilderness areas, or pestering the hell out of other people to sign petitions about this goddamn problem and that goddamn problem. And then there's the black problem, and the chicano problem, and the Chinese problem, and the oil spill in Alaska problem. I just can't figure it. We've only got a few short years to live on this planet, so why waste them worrying? What do you think, Mary? Are you sure you don't want to pop out for a quick drink somewhere? It might be fun.'

Bernard wasn't looking at Mary. He was gazing anxiously over her head, as if at some impending philosophical crisis. 'I know you got to get up early for work, like you said. But maybe a nice quiet drink somewhere will help you sleep?'

Mary was lying on her side on the sofa, her knees and elbow positioned to prevent Bernard from taking the seat beside her. Bernard had promised to leave in 'just a sec' three times already. Mary had bought a brand-new paperback diet and health book that afternoon at the supermarket, and had been looking forward to itemizing her week's meal and exercise regimen. Now, after nearly an hour of Bernard, she was so tired she could hardly keep her eyes open, and so sick of looking at him that she couldn't see straight.

Bernard was shifting tensely on the edge of his chair, glancing at the cold coffee cup between his hands as if it might tell on him.

'Bernard, I'm not in the habit of being impolite to my guests,' Mary said finally, and swung to an upright sitting position on the sofa. 'And I certainly don't want to hurt your feelings or anything. But I'm extraordinarily tired right now, and I don't think I'm getting through to you. I don't want to see you anymore, Bernard. And I don't mean just tonight, either. I mean *ever.*'

More abruptly than Mary could have imagined in her wildest dreams, Bernard stopped writing, calling, and dropping by her apartment at odd hours for coffee. He no longer insinuated his rather lurid compliments about Mary's vaguely revealing clothes into their aimless, one-sided conversations about basketball and bad transmissions, or affected his awkward sexual innuendoes at the most poorly timed opportunities. He no longer brushed his dandruff all over the sofa, escorted Mary to budget steak and fast-food restaurants, or forced her to watch reruns of bad seventies made-for-TV movies rather than drive all the way to the video rental store three blocks away. He no longer looked hurt and bent-out-of-shape whenever Mary rejected his advances, or tried to bribe Mary into repairing buttons and frays on his jackets in exchange for special discount coupon booklets he had received as gratuities in an advertising promotion at work. At first, there was something about Bernard's absence that filled Mary's apartment with an enveloping sense of sinister harmony. After a while, however, Bernard's absence simply enveloped.

Whenever Mary was caught in traffic on the freeway or riding department store escalators, she glimpsed people in the distance who looked a great deal like Bernard, though when she finally caught up with them they didn't resemble him in the slightest. In supermarkets she heard voices remotely resembling Bernard's in other, parallel aisles, and even the CBS *Evening News* with Dan Rather conducted interviews with random celebrities and politicians who sounded just as candidly stupid as Bernard ever had. In fact, the more Mary thought about it, the entire world seemed to be growing increasingly more Bernard-like every day. It was as if Bernard's voice was in everything now – pavements, architecture, governments, economy and foreign trade. His voice was even in the resounding silence that filled Mary's apartment every evening when she returned home from work.

'Oh holy-*leap*ing-mother-of-Christ-on-a-Bicycle,' Bernard's voice said whenever Mary returned home dressed for the office. 'Take a look at that hot number just came through the door. Over here, sweetheart. Over here, you hot number, you.' Mary knew it was crazy, but she could even hear Bernard's scaly eczematic hand emphatically thumping the plump sofa-cushion. When she looked at the sofa, however, she only saw yesterday's newspaper, or that morning's unfinished Diet Cream Soda. Then Mary went into the kitchen and began activating her various cooking appliances. While the apartment filled with this intimate mechanical chatter, Bernard's voice gradually shook apart and discorded, like receding signals on a car radio. After dinner, Mary watched TV, or played cassette tapes. She called Susan and talked for hours about unprofessional executives at work and suspicious glances exchanged between notorious couples at office parties. It was very difficult, however, for Mary to spend even a few minutes alone without thinking of Bernard.

The instant Bernard answered the phone Mary always hung up, feeling out of breath and a little hasty. But one day the following March Mary braced herself with half a bottle of California burgundy and arranged dinner at her apartment for a neutral-sounding weekday evening. Mary prepared what she considered a sexually-ambiguous meal of rice, peas, macaroni and smoked ham, and played nothing but Handel's *Messiah* on her cassette stereo. Bernard arrived outside in his car half an hour early and parked at the curb, combing his hair and examining his teeth and gums in the rear-view mirror for nearly twenty minutes before he brought his candy and flowers to Mary's door.

'If you stopped embarrassing me all the time by telling me what a hot-looking babe I am and everything,' Mary said, as gently as she could, 'then maybe I could relax more when you were around. It's not that I don't care about you, Bernard. It's just that I can't relax for one minute when you're around.'

Bernard was sitting all clenched up on the sofa, his hands pressed between his knees. He was wearing a reasonably clean and pressed brown knit jacket, tie and slacks. His recently permed hair sat atop his head like a wig or affectation. He kept nodding and agreeing with everything Mary said.

'I know, I know,' he said. 'I know I shouldn't. I understand how it makes you feel. I really do.'

As Mary talked and Bernard all too rapidly agreed, Mary felt her own firmness beginning to dissolve in the air like a lozenge in a well. Looking at Bernard, she couldn't help thinking how much of the fight had gone out of him. He must have had a really lousy couple of months, Mary thought, and felt a tidal rouse of maternal tenderness lift into her heart and face, like the feeling she always had whenever she held her sister's baby. Later, the tenderness receded after Bernard politely, almost demurely kissed her goodnight. Mary felt terribly sad and alone, as if she had been washed up by her own emotions on some black, mapless shore.

Within a matter of days, she and Bernard began dating again. Wherever they went together, Bernard was always hurrying to open doors and pull out chairs for Mary and 'discover' unusual candlelit restaurants along the beach and in the canyons. After a few months, they spent the weekend together at Mary's apartment, and the following spring made their wedding preparations for June, when Mary's parents were already scheduled to be in town.

Some mornings at six, five, or even four a.m. Mary awoke in her bed with a sudden anxiety rush, jarred and breathless, as if some manic child underneath the floorboards had begun pounding an untuned piano. Bernard tried to soothe her with soft caresses against her back and legs, or even a small glass of brandy from the kitchen.

Bernard hardly ever said anything anymore, usually because he was too busy entertaining requests. 'Is there anything else you need?' he might ask. 'What would you like to do today? Would you like to go out to dinner and see a movie? Or do you need your own space again? Would you like me to get out of your hair and leave you alone?' When Mary fell silent, Bernard fell silent too. They would sit in bed together, or at the breakfast table, and feel the silence settle in around them like a sort of abstract tenant. After a while, Bernard would pick up the paper, and Mary would go into the bathroom for her morning shower.

Mary tried to explain to Bernard, but she couldn't seem to get through to him.

'You know, Bernard. You don't have to do everything I tell you all the time. You can make some decisions for yourself, you know.'

'I don't understand,' Bernard would say, wincing with confusion. Tears sprang to his eyes. 'Do you mean you want me to *not* do what you tell me?'

Almost overnight, Bernard was thoughtful, sensitive and considerate; he even began entertaining more liberal opinions about child-care, environmental pollution, and low-cost housing. He was as insulted and enraged as any woman by Mary and Susan's stories of sex-discrimination at work, and insisted on sharing all household chores equally, even when Mary wanted to wash up alone in the kitchen because it helped her relax. Bernard was filling their home with soft, unyielding pressure. Mary could feel the weight of him every time she entered the house, even when he wasn't there. She began casually perusing the pages of glossy travel magazines. Every few weeks or so she considered having her hair restyled. It seemed as if life just went on and on, senseless and resolute, whether Mary wanted it to or not.

Then, in the second year of their marriage, Mary became pregnant. She had been on the pill, but she became pregnant anyway.

Late one night, as if in a dream, Bernard became Bernard again. There was something dry, immense and unavoidable about him, like a landmark in the desert. It was as if Mary had been looking for him all along, but had forgotten what she was looking for until this sudden moment when she accidentally found him again.

'Babes dig me,' Bernard said, drinking Colt Stout Malt Liquor from a sixteen-ounce can. 'Like, I don't mean there aren't plenty of babes that *don't* dig me, understand. I'm just saying that when chicks *do* dig me, they *really* dig me.' Bernard had tweezered something from inside his left ear with thumb and forefinger. After giving it a careful inspection, he placed it casually on the rim of his empty coffee saucer. 'I know it's not like I'm that good *looking* or anything. It's more like a sort of molecular attraction; it's got something to do with biochemistry and physics. I meet a certain type of babe, you know, and suddenly she's all over me. She's calling me all the time, day and night, night and day. Before I know it, she's introducing me to her family, trying to move her stuff into my place, and getting all jealous whenever I even *look* at another woman. I always try to let them down easy, you know. But sometimes that can be pretty difficult, as I'm sure you understand.'

Bernard was sitting on an unidentifiable sofa in an unidentifiable room somewhere. He seemed very far away, but his weight and voice were imminent, as if Mary were viewing him through the wrong end of a telescope. As Mary stepped closer, Bernard went further away. As Mary stepped further away, Bernard moved closer. There was a spot of shaving cream under his left ear. Then, as if Mary had no control over anything anymore, she woke up.

'What's the matter, honey? Bad dream? Did you have another of your bad dreams or something?'

Mary was still trembling. She wrapped her arms tight around her stomach. She could feel the resonance of life down there. Fluids were being redirected. Strange intentions assembling like armies on a battlefield.

'What's the matter, honey? Tell me what's the matter and I'll make it better.' Bernard was leaning into her. His weight was vast and hot and dense.

'I don't know,' Mary said. 'I don't know what's the matter. I wish I did. I wish I wasn't the only one who never knows what the matter is. I wish I wasn't the only one who never has a single clue.'

And then, when there was nothing she could do or say to prevent it, the real Bernard moved closer and wrapped his large arms protectively around her.

DIDN'T SHE KNOW

Alison Parrott was the sort of woman who appealed to old men. 'I never ask them for anything,' Alison told her landlady, Mrs Flanders, one sunny afternoon in Canoga Park. 'But they give me stuff anyway. Nice stuff, usually; sometimes even pretty expensive-looking stuff. Stuff they don't need, but stuff they're too fond of to ever throw away.' Rattly ostentatious bracelets and watches, countertop kitchen appliances, book-club editions of forgotten lurid best-sellers, sun-stained draperies and discontinued grooming supplies. 'Most of their wives died many years ago, and they tell me many sad stories about them. I'm really surprised, for example, just how many elderly women have died in motel bathrooms and public laundromats. You'd be surprised, Mrs Flanders. There's been quite a few of them, actually.' Mrs Flanders, meanwhile, sniffed suspiciously at the dusty, vaguely ominous gifts and cards that gathered daily in Alison's studio apartment like errant premonitions. A few months later, Alison purchased her own condo in the San Gabriel mountains. She leased a silver Mazda RX-7 and disdained fast-food. She continued working as a waitress at the Glendale Coco's, but she no longer worked nights or weekends. She never saw Mrs Flanders again.

Usually they arrived in the early morning, about the same time as the newspapers. They sat at the counter beside the cash register and requested buttered toast, coffee, sweet rolls, English muffins, tea with lemon, foil-wrapped after-dinner mints and repeated directions to the restroom. Alison, meanwhile, provided them with many easy, complimentary attentions. Bright smiles, unsolicited refills, additional units of jelly and jam; occasionally she even unfastened a liberal button of her striped uniform blouse. Sometimes the old men sat there for hours, patting their thin hair in place and humming distantly while they pretended not to look at her. They seemed strangely collapsed and inconspicuous inside their faded slacks, iron-stained shirts and loosening, inelastic white wool socks. They said things like, 'Very good coffee, Alison.' 'Always surrounded by your boyfriends – aren't you, Alison?' 'I really must be going – but okay. Just fill it halfway, Alison. And then I really must be going.'

Whenever Alison's young men dropped by for coffee, they and the old men grew moody and indecorous with one another. The old men didn't approve of Alison's young men, who often wore heavy leather tool belts, T-shirts and Levi's, and displayed their muscular chests and tattooed biceps while they smoked cigarettes and scratched their lean, sunburnt stomachs. 'If only you could see how silly you look,' Alison admonished her young men whenever they grew excessively curt or impolite. 'Being jealous about some sweet old man who hardly has any friends left alive in the whole wide world, and who just happens to leave me a really nice tip every once in a while. I get off at three, honey, but I won't be able to meet you until at least half past eight. Old Mr Saunders is at St Vincent's again for another of his hernia operations, and I promised I'd drop by in case he needed anything.'

Hospitals were echoing, angular white places where incoherent elderly women perambulated about with the aid of aluminum walkers, and everything smelled faintly of chemistry and soiled cotton. Middle-aged nurses hurried from room to room transporting bright, color-coded dosages of medication in tiny corrugated paper cups. Black orderlies parked their empty gurneys against bare beige walls in order to smoke their generic menthol cigarettes and watch Alison walk by. Alison always wore bright wool skirts, glossy lipsticks, and sheer silk blouses. Even the high harsh fluorescents could not imperil her cautiously rouged complexion. Alison liked the high ceilings, and the sense of close, distantly articulated corridors enclosing and defining the thick spaces all around her. Most of all, though, Alison liked the sound of her own heels clacking across the antiseptically swabbed gray linoleum floors, and the sight of her reflection gliding imperturbably across the round surfaces of chrome wheelchairs and meal trays like some sort of transistorized ghost.

'Of course, you'll be up and around again in no time,' Alison assured Mr Saunders. 'In fact, if I do say so myself, you're starting to look one hell of a lot better already – and I'm not just saying that to make you feel good. You're definitely looking a whole lot better than you did the last time I came to visit.' Mr Saunders lay motionless beneath his excessively starched sheets and didn't say anything, his cheeks deflated and sallow, his gummy mouth wide open and long deprived of teeth. He breathed stiffly and with visible effort. In fact, Alison was beginning to think that breathing was the only visible effort Mr Saunders was capable of making anymore.

'I think you'll have to admit, Mr Saunders,' Alison said, 'that life can be a very special experience indeed. As long as there is life, there is hope – isn't that how the old saying goes? If they can put men on the moon, then I certainly don't see why they can't cure some silly little intestinal disorder like you've been experiencing lately – and I don't care *what* those overpriced doctors say about your colon. Mr Saunders? Did you want to rest? Am I talking your ears off? Am I boring you half silly?' Alison took Mr Saunders's hand in her lap. The hand was faintly and placidly warm, like the tiny engine of a portable cassette player. Alison held Mr Saunders's hand in her lap for nearly ten minutes before she left, but Mr Saunders never seemed to notice.

The funerals were always held at sparsely attended veterans' and welfare cemeteries. Often military guns were fired, and American flags folded atop the clean caskets before they were deposited into deep, rectangular pits – in order to keep the patriotic old men warm, Alison liked to reassure herself. Down there in the deep somatic earth, the old men recovered warmth again; they regained the ancient communities of families and loved ones they had lost to wars, disease, and time, communicating with one another through intricate, resonant folds in the planet's subcrustal blankets and mantle. Perhaps, down there, they enjoyed spectacular transformations. Their frayed parched skin and muscles burst forth new wings and shiny, segmented bodies. They burrowed passages and wide regressive chambers, long ornate halls decorated with gold and bronze medallions, elaborate tapestries and muted brocade carpets. They gathered together in the dark buzzing colonies of their past and discussed remote details of their lives while the earth reminded them of generalities far more ancient. The beat and the hum, the spin and the heat.

After the ceremonies concluded, Alison was ritually approached by other, even more attentive old men. These old men wore ill-fitting military uniforms and bright deployments of citation and medal. It was as if every funeral articulated Alison with some secret network of old age and dark assignation. 'Well, sure. I don't see why not,' Alison told them, applying rouge to her cheeks with a bright flashing compact, watching herself take a breath mint with her flat red tongue. 'I don't see how one little drink could hurt anybody. And, of course, as you well know, Mr Saunders always told me so many nice things about you.'

Mr O'Connor and Mr Cherboninski, Mr Jones and Mr Reilly. Eventually Alison purchased a four-bedroom modular-style home in Pasadena, as well as a fully winterized cabin at Lake Arrowhead. She sublet her condo to graduate students from Caltech and, occasionally, even to some of her own young men. Young men were very difficult, peripatetic sorts of people who were always leaving and being replaced by other, even younger men. All day they brooded alone in their hot rooms, preparing themselves sullen meals of canned and frozen food. Their brains were thick, imprecise objects that did not grasp things so much as obtrude clumsily against them. The young men always told Alison to return home quickly. But then, whenever Alison did return home, they greeted her with resentment and hard looks.

'These days, young men are filled with terrible anger,' Alison told her only true, abiding friends, those faded and absently remembering old men. 'It's as if youth were a sort of fury, something rapacious and excessive, blundering and unprofound. Young men never do anything constructive, and all they ever think about is making love to young women. Young men aren't content to be planted in the warm earth. They don't want to make themselves plentiful or productive in any way. They want to take beauty, and keep it, and hurt it because young men are eternally doomed to love only their own youth. Until, that is, they don't have it anymore.'

In July Alison was taken to court by Mr Reilly's estranged widow, Madeline, as well as Madeline's two grim and mercenary foster children, Oswald and Jeannette. 'Roger Reilly and I enjoyed nearly forty-four years of happy marriage,' Madeline Reilly told the courtroom, 'until *Miss* Parrott came along one day and ruined *everything*. Suddenly, Roger began to spend an inordinate amount of time in the bathroom. He began combing his hair with some sort of slick, greasy substance. He made frequent trips to Walgreen's and Long's Drugs, where he purchased unusual creams and ointments, vitamins and skin conditioners. One day while I was baking I nearly dropped all my cookies when Roger returned home wearing some fruity men's cologne. I should have seen it coming. I should have suspected something was up. It wasn't until two weeks after he died that I learned about the will.'

Mrs Reilly's voice grew increasingly dry and severe. Alison, meanwhile, paid many polite covert attentions to individual members of the jury with her eyes. She exposed one knee and a lustrous glimmer of nylon; then, with a barely perceptible flourish, she crossed it over her other knee. On the final day of the trial, the judge requested a private counsel with Alison in his office. The judge had beautiful white hair, a Thunderbird Classic convertible, and a large, generally unfrequented home in Glendale. Eventually Alison began visiting the judge's house every Thursday afternoon, whenever the judge's wife was out working for one of her local school charities.

'It's not like I don't understand Mrs Reilly's point of view,' Alison told the judge one warm autumn evening many weeks after the trial's conclusion. 'Just because I don't happen to agree with someone doesn't mean I can't understand their point of view. But that also doesn't mean I can just disregard Mr Reilly's last wishes, either.' While Alison talked, the judge held her hand and dreamed of remote, gliding airplanes and beautiful young women.

Although Alison considered herself a happy and firmly centered individual, some nights she started suddenly awake in her bed. She sat up in the dark bedroom and felt the breathing weight of the young man beside her, entangled by her silk sheets, enveloped by the translated warmth of her body in the blankets. Her bedroom, her house. Alison would get quietly out of bed and pull on her blue cashmere robe. The young man would sigh and turn over in his dreams of her. As Alison moved alone into the dark house she felt herself verging upon vast unlabeled places. She felt a quick thin rush of vertigo, and sensed the heavy white gaze of the moon in the garden.

Methodical spiders spun shining nets out there and swung from invisible threads. In the trees birds gathered sticks and brushy moss and bugs, and in the earth strange blind creatures burrowed and tugged and crawled. We build and build, Alison thought. We build houses, estates, families, empires, even notions about the way worlds work. We build and build, and then we all grow old and die anyway.

Earlier that day in Whittier Alison had visited her friend Mr Peabody, whose brain, heart, and lungs were attached to a battery of humming, discordant machines with round, radar-like screens and formidable steel paneling. Mr Peabody had not died exactly, but he was not altogether very well. For the rest of his life, Mr Peabody would have to lie alone in his expensive room, embraced by the steel musculature of technology, infused with the glowing and often perilous blood of strangers. Alison had worn her low-cut, slim-waisted green silk Halston blouse when she visited Mr Peabody that afternoon. The green silk Halston blouse had always been Mr Peabody's favorite.

'They all think I was just friends with Mr Peabody because I wanted his money,' Alison said out loud. Her young man continued to sleep, breathing noisily in a way that Alison always found comforting. He was very curled and very tan, and Alison liked to examine his strong back while he slept. 'But no matter what anybody says, I'll always miss Mr Peabody very much. We used to play chess together all the time, and he never once let me win on purpose. Baby? Are you awake?'

The young man stretched in his sleep, anemone-like, as if slowly unsprung by Alison's indelicate touch.

Alison touched him again.

'Come here, baby,' Alison said. 'Come over to this side of the bed where it's warm.'

It was as if Alison could feel old age gathering in the world like heat or information. Old men rang her at home and on the cellular phone in her car. They left messages on her answering machine, cards and flowers on her porch. They mailed her money orders and savings certificates, stock options and government bonds. Alison's doorbell was always buzzing, and immaculately uniformed delivery personnel presented her with outrageous kiss-o-grams, personalized birthday ballads, pungent roses, and elaborately packaged European chocolates. Gift-wrapped boxes arrived bearing stereos and diamonds, kittens and puppies. 'They're all such sweethearts,' Alison said, pale and deflated on her leather-upholstered sofa, the mountains of pink-chiffon tissue paper and coiled ribbons lying collapsed around her like flushed, napping children after a sugar high. A new Guatemalan wool sweater lay draped across Alison's perfect white knees. 'I don't know how to thank them sometimes. I don't even know what to say.'

The old men suffered debilitating strokes and hemorrhages, senile dementia and gout, neglect and bad dreams. 'Have you ever thought of buying a VCR?' Alison asked Mr Samson, who always called well past midnight. Mr Samson had outlived two wives and three middle-aged children. He never slept more than one or two hours each night, and often awoke to find himself somnambulating along Lankershim Boulevard, or sitting alone in bright all-night laundromats as far away as Sepulveda. 'There's a lot of good movies available on VCR these days – even many forgotten classics from the thirties and forties.' Alison was holding a cold damp cloth against her forehead. She sat on a deeply cushioned sofa amidst the wide, empty room like a shipwreck survivor on a raft, leaning into the luminous telephone, tugging aimlessly at her new haircut. 'I know it's boring, Mr Samson,' she said. 'I know it's lonely. I know you can't sleep. I really do understand, Mr Samson. I really, really do.'

Alison stopped answering the phone and opening her mail. Every day she sat outside on the sunny back porch drinking tall, icy glasses of mineral water with lime, eating fresh fruit salads, and listening to baroque music on her portable CD system. She wore sunscreens, sunglasses, and Vitamin E lotions. She did deep knee bends, leg-lifts, tummy-tenseners, and facial aerobics, examining herself in the full-length Cinerama-style dressing-room mirror about two hundred times each day. In January, she purchased an exercycle; in the spring, she radon-proofed her basement. Every night she made love with the young men until dawn when they fell asleep without her, and then wandered alone out to the back patio where she gazed at the far, Spanish-style red-brick roofs and leafy, fading verandas of the San Fernando Valley. Everything seemed perfectly immaculate and disingenuous at night. Automobiles raced along Van Nuys Boulevard, disguising the race and beat of Alison's own young heart. 'Who wants to live forever, anyway?' Alison asked herself, sipping wine coolers and smoking clove cigarettes. 'That would be really boring. You'd run out of new things to do. You'd have worn all the nice clothes already. You'd have taken all the best vacations.'

Alison could feel youth being methodically extracted from her body like the thin smooth exudated thread of a silkworm. The sleepless old men lurked outside her house at night, pacing in the dry dark street, looking very obvious in the backseats of their Yellow Cabs. Mr Brenly and Mr Chin, Mr Bradley and Mr Gray. Sometimes they ventured up the long cement path to her door and rang the bell, or knocked gently a few times. Meanwhile, Alison stood invisibly behind her wide tinted patio window, watching them. 'Why don't you just give them a good kick in the butt?' Alison's young men suggested, squeezing rubber wrist developers, or pumping hard black iron weights as they stood behind her in the dark living room. They wore tight silk gym trunks and bright tank-top T-shirts. 'I'll bet that's one message they'd comprehend *muy* pronto, sweetheart. A good swift kick in the raggedy old derrier sometime.' Alison took another hit from the joint and felt the fuzzy comprehension lift high into her chest, throat, and bony face. It was filling her skull and sinuses now. Now her eyes were swimming in the damp sweet mystery of it.

Eventually the old men went away again in their cabs. The young men unclenched their exercise equipment and shrugged their way into the kitchen for glasses of mineral water and light beer. 'If I were an old geezer like that, I'd be too embarrassed to go around hassling some sweet young thing that didn't want to be hassled,' the young men muttered into their sparkling Perrier, tearing celery from crisp stalks. Then they spread peanut butter, chives, and sour cream on the celery with a blunt, flat knife. 'And, if I couldn't live with it, man, like I'd just go off quietly and cut my own throat, you know? I'd jump out a twelve-storey window – no parachute. I'd fire a Teflon bullet point-blank right into the old brain-basket – that's what *I'd* do, sweetheart. Believe you goddamn me. You wouldn't catch *me* hanging around for one goddamn minute where *I* wasn't wanted. And that's for *damn* sure.'

Even at night while she slept, Alison could feel the old men going away from her. They were departing on trains and buses and planes. They were hiring private cars and drivers, or gliding across misty foreign oceans on hissing hovercraft and wide slow exorbitant luxury liners. Usually they took care to shave and comb their hair first. They wore their best pressed white shirts and unsnagged smooth cotton trousers. They listened to their favorite records on the phonograph, and thumbed through old leather-bound photo albums before replacing them on high, dusty shelves in neglected closets and outdoor storage sheds. Then they latched all the doors and windows in their memento-cluttered rooms and houses. They never carried more than a single piece of hand luggage, or perhaps a few spare shorts and socks in roughened brown-paper shopping bags. The old men were going away, and Alison knew she would never see any of them ever again.

Alison attended their funerals incognito now. She wore long black dresses and hats and ominous-looking black satin veils. She sat anonymously in the back row while the old men watched their compatriots being lowered into the warm earth. Sometimes the mournful old men glanced quickly and surreptitiously at Alison over their shoulders. They fidgeted on their clumsy metal folding chairs, or coughed awkwardly into their red, arthritic fists. Most of the time, though, they worked very hard not to look in Alison's direction at all. Sometimes it was because they didn't recognize her. Sometimes it was because they did.

As the old men went away, the young men grew angrier and angrier. They slammed doors and cupboards, they smashed plates and windows and clocks. They drove Alison's cars out on the freeway at night and crashed them, and returned home smelling of petrol, gin and imprecation. They made abrupt, sore love to Alison in the middle of the night, usually while she was still half-asleep, and when they left her now they took things. Money from her wallet, credit cards, checks; jewelry, stereo equipment, cameras, and linen. Often at night Alison could hear the young men rummaging through the house while she pretended to sleep. They knocked over lamps, they collided with the abrupt edges of tables and sofas, they overturned bookshelves and glass-paneled cabinets and stuffed chairs. 'The bitch,' they muttered in the dark, as if they could actually smell the old men still lurking somewhere in the fabric of Alison's life. 'The lying little bitch,' they said. 'It's around here somewhere. There's more. It's here. There's more around here somewhere.'

In the mornings they were gone, and Alison tried to make amends. She replaced unbroken dishes on their shelves, and soothed the scratched furniture with conditioners, waxes, and polishes. 'You can't hate a young man just because he's angry,' Alison said out loud to herself in the large, echoing kitchen. She was sweeping motes and shards of brilliant crystal into a tidy aluminum dustpan. 'Young men live at a different pitch, a different frequency from young women. Sometimes I think young men must be the loneliest people on the face of the earth. Other times, though, I think the loneliest people on the face of the earth have got to be young women.'

The old men were almost gone now, and already Alison was beginning to miss them. She could remember the way they breathed, and their white fingers trembling against the rims of regulation-size Coco's coffee cups. She remembered the linty, crumpled bills and excessive pools of change they left behind, the smell of lanolin and Vaseline, and the way they watched her when they thought she wasn't watching them. Some nights she would drive out to the San Fernando Valley, or down Highway 1 to Retirement Village, and coast slowly past their untended ranch-style homes, which were filled with watery silence and ripe, abandoned equity. Sometimes she would pull over for a little while and gaze at the dry, unraveling yards, the alert Real Estate signs, the dark glimmering windows from which all the curtains had been removed. Once the old men were gone, they could never come back, and this thought filled Alison with both a sense of profound sadness, and a thin hormonal rush of visceral relief.

Alison couldn't sleep at night anymore, the cold was so bright and unremitting. It penetrated walls, insulation, clothing doors. It radiated from televisions, kitchen appliances, porcelain bathroom tiles, the flat moon and glittering, inalienable stars. Down there in the dirt and the earth, the old men whispered. 'Alison,' they whispered. 'Was it something we said? Was it the way we looked at you, or was it the way we always tried to touch your hair?' Alison couldn't explain. She turned over and over again, unable to sleep. 'It wasn't you,' Alison tried to tell them. 'It's something wrong with *me.* I don't even know how to explain it. But it was never anything at all the matter with *you.*' Sometimes she felt as haunted by the old men now as she did when they were still alive.

Meanwhile, outside her home, the young men arrived drunk and unscheduled, confronting one another in one another's stray clothing. They cursed and battled across the yards and driveways, overturning birdbaths and garbage cans, mailboxes and concrete lawn ornaments, pounding at one another's hard chests and faces and stomachs until neighbors called the police. There was all this noise and clatter in the world, Alison thought, feeling the cold center of her heart where the old men whispered. All this roil and struggle, and nowhere to just lie down, relax for a minute, and hear yourself think. The cold in Alison's body was like a recollection, a river, a simple animal notion of the heart. It circulated in Alison's blood like plasma. It brought a strange caress to the rubbery, secret interior of veins and arteries. 'You fuck!' the young men cursed each other outside. 'You stupid fuck! Who're *you* calling a stupid fuck, you stupid fuck!'

'I like boys,' Alison told the old men, the only men she had ever known who listened. 'The trouble is, I've never been able to tell whether boys liked me or not. I mean, sometimes they can be quite sweet, really. But other times it's like, well, I don't know. It's like I'm everything they don't want. It's like I'm everything about the world that they don't love, and that doesn't love them.' Every morning Alison sat among the ruins of her living room, drinking instant coffee and trying to explain. Throughout her house the young men had ripped the upholstery from sofas and chairs, uprooted the pile carpet with chopping knives and fireplace instruments, and torn all the framed landscapes and photographs from the walls. 'It's sort of the same way I felt about you,' she told the old men, and at this point she started to cry. 'It's sort of the exact same way I treated you. Though I didn't mean it. Because I didn't *mean* to treat you that bad. I just didn't know I was doing it at the time.'

At night, in her bed, Alison lay quiet and listened for sounds from the driveway, the garden steps, the porch. The young men were stealthy and compact. They slammed their car doors and cursed. They were here. It was tonight. They wanted it now. They wanted everything tonight.

They kicked cans in the street. Sometimes they howled like wild dogs and threw their bottles at passing cars that honked back at them, as if they were engaged in a primitive debate. Then they let themselves in Alison's front door with their key. Alison usually knew his name before he was halfway down the hall. Bobby, she thought. Or Reggie, or Stan. Down there in the dark earth, the old men continued to whisper, as if they didn't know they could never return. 'They don't love you,' the old men said. 'Only we really love you, Alison. And just look what you did to us in return.'

Now the young man was prowling through the office, the guest room, the closets, the bath. Lights were snapping on and off, blankets and mattresses being whipped from beds, lamps crashing insensibly to the floor. Then the young man was in the hallway again. Then, covered with darkness, the young man was standing in Alison's bedroom again.

'It's okay, baby,' Alison said. She was clutching her blankets and sheets in her cold hands. Everything she touched these days was cold. 'Just relax, baby. Don't be mad at me.'

She could feel the heat and steam of him expanding in the room, lost, dark, limbic, and round. She could feel the ridged scars along the spine of his forearm, his bruised lip, the tiny weblike cracks in his face and his skull. She could feel his hot fists clenching, unclenching. The young man wanted more. He wanted more, but no matter how many times she tried, Alison couldn't seem to explain.

'They're all gone,' Alison whispered. 'All buried and dead. Did you hear me? They're all gone, sweetheart. Why don't you come lie down with me and get warm.'

This was night and recollection all at once. The young man moved closer, darkness sliding across him like a sort of film. The young man kept moving closer. This happened to Alison night after night after night. Suddenly, Alison's world was filled with nothing but young men.

'It should've been you,' the young man said, coming closer, not starting to reach out for her just yet. 'You bitch. It should have been you. It should have been you.'

IN THE TIME OF THE GREAT DYING

When I was fifteen and my brother eleven, our mother began undergoing a series of surgeries and treatments that left her bedridden for weeks and months at a time. At first they removed a portion of her left foot, and then her entire left breast. They removed her cervix, womb and fallopian tubes. When I received my trainee driver's license at fifteen and a half, I regularly drove our mother to radiation treatments at Children's Hospital in San Francisco, where I sat in the lobby and waited patiently among assortments of other radiotherapy patients, hurrying nurses and vague, irreproachable doctors. Like our mother, the other patients arrived wearing pajamas, housecoats and slippers. They displayed raw, reddening tumors on their throats and faces. Sometimes they were missing teeth, ears, eyebrows, limbs and noses. They seemed slightly stunned and nondescript, as if they had either forgotten everything they had ever known, or were suddenly remembering everything they could never forget.

Whenever our mother was in the hospital, my brother and I would meet after school and ride a sequence of infrequent buses to visit her. As we left our wide suburb, the world grew closer and more complicated. Loud and abusive drunks joined us on the bus, negroes, hispanics, and other kids our own age who always looked a lot more formidable than we could even pretend to be. I often imagined scenes in which I protected my younger brother from attacks by these people, countering their strength and firepower with my own superior brand of wily ingenuity. My brother, however, always maintained a curious attitude of total serenity whatever our surrounding circumstances. While we rode along he gazed out the window with that dully reflective, disconcerned expression I always admired in him but could never emulate. I, meanwhile, nervously mapped out our itineraries.

'We'll get off at Stonestown and catch the 28,' I said. I was jiggling my knees and sucking a Life Saver. 'Then we'll take the 46 straight up Geary to the hospital. We'll see Mom, and eat dinner in the cafeteria. I think *I'm* going to have lasagna, just like yesterday. Then we'll catch the bus home in time to watch *Mod Squad.* I think it's going to be a particularly good *Mod Squad* this week. It's about these bikers who take over a small town, and the gang leader of the bikers turns out to be Link's older brother.'

My brother didn't even look at me sometimes, but I could see his face reflected in the window. He was gazing out at the city as if it were some wearisome mirage. 'Maybe you could just shut up for a few minutes,' my brother told me softly, wearing his congenital and abstract scowl. 'Why don't you put a lid on it sometimes. Why don't you stop jiggling all the time and try to relax.'

They removed her spleen in July, and by the time our mother returned home from the hospital she had already begun losing weight. My brother prepared her meals while I gave her back rubs and performed adjustments on the new color television she had recently purchased herself from Sears. By this time, she was dosing herself with a complex medley of pain-killers. They had given up on chemotherapy two months before the removal of her spleen. They had discontinued radiotherapy six weeks before that.

'Our mother's going to die soon,' I told my brother one afternoon. We were sitting in the living room, and he was watching *Hercules in Hell* on Channel 44's Afternoon Movie Playhouse. 'We might as well get used to the idea, and start discussing how we'll handle things after she's gone. Maybe it's too early to come up with any definite answers or anything, but we should at least start talking about it. Just to be realistic. Just so we know what to expect.'

'She's not going to die,' my brother said. 'She's just really sick. She works too hard and needs a little rest, that's all.'

Hercules, portrayed by my brother's favorite actor, Steve Reeves, was chopping the heads off a Hydra with his sword.

'Our mother hasn't been to work in over thirteen months,' I told my brother. 'Most of her bodily organs have either been removed or infected by orange-sized tumors. We're living off Blue Cross and Social Security. It's about time we faced facts.'

'I think you're a big asshole,' my brother said, opening his Civic Studies workbook. Then he drew a pair of enormous breasts on Abraham Lincoln, and a long urinating penis on a statue of George Washington.

'In the time of the great dying,' I told my brother, 'all the dinosaurs disappeared at once from the surface of our planet. The Tyrannosaurus Rex, for example, the world's first meat-eater. The bony Stegosaurus, which had a brain about the size of a peanut. The Brontosaurus and Ichthyosaurus, Pterodactyl and Saber-toothed Tiger. There are many interesting theories about why the dinosaurs vanished so suddenly, but the most interesting theory, I believe, is the one about a gigantic comet which crashed into the planet Earth, sending up thick black layers of ash which disrupted normal weather patterns for hundreds of centuries. Or perhaps the sunspots did it. Or a terrible disease. Or massive volcanic explosions. Or perhaps the dinosaurs were taken away by unbelievably big spaceships to interstellar zoos in the Orion Nebula. Or perhaps the dinosaurs didn't really die at all. Perhaps that's what's hiding at the bottom of Loch Ness in Scotland, a great sea-breathing dinosaur. It's possible, you know. Almost anything's possible.'

'Anything *is* possible,' my brother said. We were waiting downstairs in the lobby of St Jude's Hospital. Our father was coming to pick us up and drive us to his apartment in Marin, where he taught Composition at the City College and dated a succession of pretty, secretarial-types who took a cautious, even-handed interest in both my brother and myself. 'I think she's looking a lot better today, don't you? I think this may have been it. I think she's going to start feeling a lot better from now on.'

'I've always been sort of an expert on dinosaurs,' I told my brother. 'Mainly on account of I've always been especially interested in things that happened in either the super-distant past, or the super-distant future.'

*

My brother and I continued going to school each morning, and returning home each afternoon. We did our homework, observed television, washed dishes and vacuumed the rug. I forged our mother's signature and cashed her checks at the bank, depositing her income from Blue Cross and Medi-Cal.

'If she was going to die,' my brother reasoned, 'then the nurse would have told us. Or her doctors would've come by.'

'Maybe they've forgotten about her,' I said. 'Like the way sometimes we forget about her. Sometimes, you know, I keep forgetting she's here. I keep thinking she's still back at the hospital.'

'Somebody would have told us something,' my brother said. 'They wouldn't just leave us sitting around with a dying mother on our hands, would they? Would they?'

'Maybe they did tell us.' I was gazing out the venetian blinds at the rolling Pacific fog. Then I looked at my brother. His clothes were excessively rumpled, his eyes shadowy and bruised from insufficient sleep. 'The time before her last operation. Didn't she say something about six months? How long ago was that?'

'I don't remember anything like that,' my brother said. Then he gave me a sudden push. 'You're just making that up. You're just making that up to scare me.'

'Don't push me,' I said.

'You're just making that up.' My brother was shouting now. All the calm indurate conviction I always admired in him suddenly vanished. 'You're making it all up just to scare me. You want to pretend you know everything, but you don't know *any*thing.' He pushed me again. 'But you don't know anything. You don't. You don't really know anything.'

'Don't push me,' I said.

'You don't know anything,' he said. 'Just shut up. Just shut up and leave me alone.'

I tried to grab his hands but they twisted away from me.

'Don't push me,' I said. And then he pushed me again and I pushed him, and he fell against the sofa springing up again instantaneously, as if he would resist gravity too.

'You can't do this to me!' He was screaming and crying. I couldn't even recognize his face anymore. 'You have to stop! You have to leave me alone!'

And then I pushed him again, and again. And then I pushed him again and made him stop.

Perhaps in the same strange way I can never forget our mother now that she is dead, there were times during her illness when my brother and I managed to forget her altogether. My brother attended his fifth-grade classes just down the street from my high school, and on days when I was feeling particularly wired from too many cigarettes, I would go to meet him outside his classroom door. We would walk to the Serramonte shopping mall together, where I stole paperback science fiction books from the B. Dalton's, and my brother bought excessive amounts of candy and junk food which he ate with cool and intrepid abandon. We would sit on the benches beside the garish, glittery fountain and watch the shoppers pass by, notoriously mature couples from my high school, pretty girls with new cosmetics, gray and severe old women. I would show my brother my new science fiction books, and he would offer me portions of his Slim Jims, beef jerky and candy corn. I would point out the pretty girls I had a crush on. My brother would promise not to tell.

'So what are you going to be when you grow up?' I asked him.

'An astronaut,' my brother said, with blithe conviction.

'You *should* be an astronaut,' I said. 'You'll be famous and important that way. You'll travel to amazing places. You'll make more money than you'll know what to do with. But that means you have to work pretty damn hard in school. You'll have to be really good at math and science.'

'I'm already good at math and science,' my brother said. 'Math and science are my best subjects in school.'

Sometimes we might even go to the movies, or bowling in Westlake. It was always windy and cold, and I chain-smoked my cigarettes, and my brother lapsed into his formidable, arcane silences. The streets were all very familiar. My brother and I could walk for miles and miles without pausing even to check where we were, or decide where we were going. During these long walks together, I don't think I ever shut up for one minute.

'We'll go to the movies,' I said. 'Then we'll have something to eat. We'll have something with fruit and vegetables in it, since you've been eating nothing but crap all day. Then tomorrow we'll bowl three lanes – I think I've got almost five dollars. It's such a great day. I really feel good today. Tomorrow I've promised myself I'll study harder for school. I really don't think I'm living up to my potential, school-wise, at all. So what do you think about those Giants? A lot of people are saying we shouldn't have traded Bonds, but I think Bobby Murcer's a genius outfielder, don't you?'

I could talk for hours and hours, but eventually my brother's silence would start to weigh on me. I'd look at him out of the corner of my eye. I'd wonder if he was growing sick of hearing me work my mouth. On days like this, however, when we were alone and our mother was far away and completely forgotten, my brother didn't seem so abrupt or contemptuous anymore. Sometimes he even smiled a little while I talked, as if he were riding a firm, predictable current that would carry him exactly where he wanted to go. Sometimes he even joined in the conversation a little.

'Yeah,' my brother said. 'Tomorrow we'll go bowling. I think I'd really like that.'

Then we continued on through the foggy streets to our home.

GHOST GUESSED

Nor mouth had, no nor mind, expressed
What heart heard of, ghost guessed.
GERARD MANLEY HOPKINS

'Didn't expect you home so soon,' the ghost said, seated on Mother's prize rococo revival sofa, a late-Victorian design with mahogany frame and black horsehair upholstery. The ghost's features were vague, faintly phosphorescent, like mist in the beam of an outdoor film projector. 'I knew those stale pills of Mother's wouldn't do the job. I told you we should use the gun. But you wouldn't listen to me—'

Kenneth Millar shut the front door, attached the chain, turned the bolt. He knelt, removed his Hush Puppies and placed them on the plastic mat, toes square against the wall.

'What did we *buy* a gun for if we weren't going to use it? You only had to pull the trigger once, one bullet, bang, that's all she wrote. Then you wouldn't have had time to chicken out, to get on the phone to every hospital in the county . . . Hey, where you going? I'm not finished talking to you!'

In the kitchen Kenneth filled a trembling Dixie Cup with Sparklettes. He still felt weak, somewhat dizzy, nauseous. He sat and braced his elbow against the kitchen table as he drank, swallowing against a sore throat. The lavage tube had blistered his trachea, the doctor told him, and prescribed antibiotics.

'I'm not letting you ignore me.' The ghost stood in the hall.

Sunlight from the kitchen window angled through his body, illuminating a soft blizzard of dust motes. 'I knew you'd try that. It's just your style, just the cheap sort of trick you'd pull. Ignore me and maybe I'll go away. But I'm not going away, pal. You better get used to it.'

Kenneth removed a dog-eared paperback from his jacket pocket. The book's open spine was thumb-soiled, the pages curiously stained in places. The book had been deposited beside his bed the night he was admitted to County Emergency. The inside front cover had been inscribed by its anonymous donor. *Only God's love will save you. Only God's love is really real.*

'Do me a favor. Before you start reading that crap, could you turn on the TV? I seem to be pretty helpless in regard to material objects. See?' The ghost reached his arm through the kitchen wall, waved his impalpable hand. 'You realize I've been stuck in this lousy house all weekend with absolutely *nothing* to do?'

Kenneth opened his book to a random page. 'In this world a few men find happiness. They are loved by God—'

'You're a champ, Kenny. You know that? A real champ.'

'Most men never know that God is . . . that God *is*. They live like animals. They are those who live to die, and die to live again . . .' The short, precise syllables buzzed senselessly in his sinuses like flies in a tin cup. After a while he gathered his courage and looked into the hall again. The hall was empty. He took his book into the bedroom and quickly shut his door. He was afraid of making a sound. His colon twitched like a tiny snare drum. He must be imagining things, he assured himself. A stale residue of Seconal, perhaps. Authentic supernatural events were reputedly accompanied by frigid temperatures, crashing plates, conventional shrieks of alarm. All Kenneth felt was a bit of melancholy solipsism, like the night he waited alone at a downtown bus stop.

The house was quiet, and Kenneth quieter still. Might as well not tempt things. He went to the rolltop desk, quietly removed from its bottom drawer a polished walnut case with stainless-steel clasps and opened it in his lap. As a child Kenneth had asked his mother for a set of army men. The army men were sold in large fishnet sacks that hung in the window of the local toy store. They came in a number of theatrical poses, firing rifles from a crouch, hurling grenades, charging with upraised bayonets. In the afternoons, restricted to the front porch until Mother returned home, Kenneth sat and watched the neighborhood children play in the abandoned lot across the street. They balanced their plastic army men atop rocks, logs, bushes, and then, backing off a few paces, hurled small stones and sound effects at them. *Pih-chew! Pih-chew-chew-chew!* One Christmas morning Mother presented him with a small oblong box, wrapped in austere yellow paper like a festive toy coffin. She sat him in the pale living room and, with much stern ceremony, permitted him to open the card first. *Merry Christmas, Love, Mother*. 'Now you can open your gift.' Kenneth peeled away the limpid yellow tissue, unhinged the tiny cardboard coffin, disinterred the layer of soft white cotton and, beneath, the heavy lead soldier. The soldier stood at attention with fixed arms. His painted uniform and features were chipped and dull. 'I'll bet there's not another boy in town who has a soldier like that,' Mother informed him. 'It was manufactured by William Britain in the early nineteen hundreds. They were the first hollow-cast military miniatures in the smaller size. This one is a French hussar. The detail work for the period is, I am told, remarkable. That hanging at his side is a sabertache, see? Now you must be very careful. It's probably best not to remove the plastic wrapper.' Over succeeding birthdays, holidays, graduations and other dismal events, Kenneth accumulated more of the toy soldiers in his desultory walnut case. King's African Rifles, Boer, Royal Sussex Regiment, USA, Montenegrin Officer, Australian, West Indian Regiment. Each item was sealed in a tiny plastic bag, a placenta for an artifact. Unable to impute to the objects any imaginative activity, Kenneth instead learned to adopt their forlorn, attentive expressions and, on rainy afternoons, stood the soldiers on his desk and stared at their faces through the yellowing plastic, conspiring with them in inanimate silence as they awaited together the tread of Mother's car on the graveled driveway.

Kenneth closed the case and returned it to its drawer. He sat very still, listening for noise from the living room but hearing only the thermostat's hollow click. After a while he checked the gratuitous lock on the bedroom door and crept furtively into bed.

Kenneth was awakened the next morning by the blare of the television. ' – and you could win – A NEW CAR!' Appropriate audience response (applause, cheers, spontaneous squeals) shook through the thin walls of his bedroom and faintly rattled the loose ends of wallpaper. He put on his robe and slippers and went into the living room, squinting at the noise.

'Morning, champ.' The ghost was gazing vacantly at the flickering screen. 'You shoulda seen it. This dumb Puerto Rican broad didn't even know the price of a bar of soap—' His features were sharper today, more clearly defined. After a few addled moments Kenneth recognized the features as his own. Soft unwashed black hair, a goatee. Anxious, bloodshot eyes. A pale, waxy complexion.

'Could you turn that down a bit?' Kenneth asked.

'Look, I can even change channels.' The ghost reached for the selector, dialed. Static pulsed on the screen.

'But can you turn it down?'

'What?'

The telephone rang amid the resumed hysteria of the audience. Kenneth took the receiver and cupped his hand over the mouthpiece.

'I said, *turn it down*.'

'I can't hear you.' The ghost reached and decreased the volume. 'This damn set is too loud. What did you say?'

Kenneth turned his back on the sofa. 'Hello?'

'Hello? Kenny? What's all that shouting going on? What's happening over there? Are you all right?'

'Just the television, Aunt Agnes.'

'What are you doing watching television at this time of day? You should be out in your mother's garden, getting some color. And where have you been all weekend? Do you know I called twice on Saturday, *three* times Sunday?'

'You must have just missed me, Aunt Agnes. Yes, the garden. Yes, yes, I will. I know she would. Of course, you're right. I know I should have called. I'll try and remember.' Kenneth glanced idly over his shoulder as he herded Aunt Agnes's questions. The ghost held one hand high in the air. He was giving Aunt Agnes the finger.

Kenneth hung up the phone, returned to his bedroom and dressed. On his way out the front door the ghost said, 'Hey, you're gonna miss the best part. Everybody gets a chance to spin this big wheel, see? And the winner gets a billion dollars or something. Look, even that stupid Puerto Rican broad gets a shot at it.'

Kenneth arrived for work even earlier than usual at Worldco Publications. He sat alone for a while in the employee lounge, a small, drab room with a few cracked plastic chairs and a Vendomat coffee machine. He drank hot, discolored coffee, listened to the tenorless strains of Muzak and watched the clock on the wall. At ten forty-five he went upstairs to his desk in Wrong Addresses. The morning mail had already begun filtering through Computer Processing, and three subscription order rejects waited on Kenneth's desk. Kenneth consulted his desktop library of Southern California street indexes and telephone directories. When the problem was more serious than faulty spelling or incorrigible penmanship he dialed the number given on the order form.

'Hello, Mr Smead? Or is it S*n*ead? I'm calling from *Real Action Detective Stories* about the subscription you ordered?'

'I don't want one.'

'I'm not trying to *sell* you a subscription, sir. I'm trying to help you receive the subscription you already ordered—'

'My *wife* ordered the damn thing and *I* don't want it!' And then the dial tone recommenced.

Kenneth had already reprocessed a small stack of cards by the time the rest of departmental personnel began to arrive, and more rejects were being brought to his desk every minute. He glanced up reflexively when he heard the snap of heels in the aisle. Veronica passed his desk and winked significantly. 'Good morning, Ken.' Veronica glistened in the bright fluorescent office: nylons, sleek skirt, lip gloss. She sat just down the aisle from Kenneth, adjusted herself in her chair, then in her compact. She tilted the circular mirror until her eyes flicked teasingly at Kenneth. Veronica's cordiality had been a matter of public record for almost a year now, ever since the afternoon Kenneth returned from lunch to discover he had left his savings passbook on his desk.

By eleven o'clock the remaining employees were at their desks, composing reports on one another's activities. Kenneth's telephone rang.

'Mr Millar, I have a *personal* call for you,' the receptionist said. 'I've explained to your friend that the company prefers that its employees not accept *personal* calls at their desk. I hope you will keep the conversation as brief as possible.'

' – Hello, Kenny?'

Kenneth hunched and whispered into the receiver. 'I don't want you calling me here—'

'Who was that lady I just spoke with? Icy. *Brrr.*'

'What do you want?'

'Just some stuff from the store. Beer, especially. Cigarettes, potato chips, sweet rolls for breakfast – are you taking all this down?'

'Listen, Kenny. I like you. I really do. And I'd like to help you. But first you've got to stop believing that life is something dainty and pristine, like Mother's chinaware.' The ghost sat in his customary place on the sofa. The blinds in the room were growing dark. He switched on the lamp over the television, selected an apple from the hand-painted porcelain bowl and gripped it like a screwball. 'You want to know your problem, pal?' His sharp white teeth took the apple. Juice spattered his upper lip and dribbled from the corners of his mouth. 'You think you're too good for the real world. You're afraid to get dirty. So what if this broad Veronica's only after our money? Let's live a little. You saw *Zorba the Greek*.' He wiped his mouth with the back of his hand. A white shred of apple remained on the verge of his lower lip. 'Romance is just an idea somebody cooked up in order to sell more mouthwash. Me, I prefer a good time. You've just *got* to get out more, pal. You're not an old woman. Mother was an old woman, and that's why she's been dead for fifteen years.' The ghost leaned back on the sofa, lifted his feet on to the glass-topped coffee table and tossed his half-eaten apple in the general direction of the fireplace. The apple bounced soundly off the ersatz brick façade and, landing on a burgundy patch of Persian carpet, rolled over twice. 'Anyway, that's how I feel. It's just pretty hard to believe that a forty-two-year-old man has never been outside the city he was born in, has never touched a drop of booze, has never done *anything* his mother warned him against since he was age *one*, has never even been—'

'Shut up!' Kenneth flung his newspaper to the floor and started up from the art nouveau side chair, Karpen and Brothers, Chicago and New York, circa 1900, listed as item number 76 in the Orange County Auction Yearbook.

The ghost deliberated, sucked something loose from his teeth. 'Kissed,' he concluded, and flicked Kenneth a sharp glance.

Kenneth stooped and retrieved the apple. The exposed meat of the apple had already turned brown in spots.

'While you're up, could you get me a beer?'

Kenneth dropped the apple in the kitchen trash and returned to the living room with a damp sponge, dabbed at moist stains on the exorbitant carpet.

'Maybe I'm talking to myself. Maybe I just like to hear myself talk.'

Kenneth got up from the floor. 'The beer is in the refrigerator, if you'd care to look for yourself.'

'Oh, great, *great*. Now you're even starting to *talk* like an old broad.'

Kenneth rinsed the sponge in the sink for ten or fifteen minutes, wringing it mercilessly. His ears were flushed and hot, his eyes stung. He heard the refrigerator door open behind him. Cans clattered dully as one was yanked free from the plastic spine.

'Scrub those dishes, Kenny, old pal. And keep practicing. You're gonna make somebody a terrific grandmother!'

When Kenneth awoke the next morning it was twenty past ten. He jumped out of bed and grabbed the alarm clock. The alarm lever had been shut off. Kenneth had his trousers hiked halfway up his legs when he realized there wasn't time. Even if he caught the ten forty-five he would still be more than a half hour late. He lay back in bed and pulled the gloomy coverlet up to his chin. He felt abysmal, sluggish. His brain and limbs felt hollow, as if he had suffered a slow blood leak during the night. He lay and watched the white stippled ceiling, and after a few minutes noticed the silence in the house. The silence was like an immense, solid object, a statue in the park. He got up and went into the bathroom, kitchen, living room. The house was empty. He lay down on the sofa. The carpet was littered with crushed aluminum beer cans, beef jerky and candy wrappers, and a heavy spray of yellow potato chips, like the fallout from some detonated mini-mart. An empty beer can was stuck to the surface of the Renaissance revival side table. Kenneth gripped the can and yanked it free, uprooting a ring of rosewood and varnish. He turned on the television, nibbled stale chips from the depleted bag.

'It must be a real thrill being on the country's number one prime-time show.'

'Oh yes, Mike, of course it is. And Lucinda – that's the character I play – has always liked to think of herself as Number One.'

Kenneth could not sleep that night. He was thinking that when Mother died he must have been twenty-seven. The ceremony was informal and closed-casket, preparatory to cremation. Kenneth sat in the front row with Aunt Agnes. The coffin was polished and imposing, like a baby grand piano. Numerous relatives were in attendance, reminiscent of Mother's album of faded snapshots. After the eulogy, strange men shook Kenneth's hand and introduced themselves as Mother's uncles, cousins and business associates. 'Let's see . . . if I'm your mother's cousin, then that makes me your . . . *not* your uncle. Your *second* cousin?' A few old women in spotty furs inquired about the estate, but the will proved ironclad. Aunt Agnes received the business, Kenneth the house and the savings account. Kenneth also inherited, as a sort of gratuity, a firm memory of Mother. Mother was not really gone. She was not reducible to dust. She would be with Kenneth always, in the back of his mind, sealed and suspended in an impervious block of clear Lucite, like a scorpion in a paperweight. Kenneth rolled over on to his side, then on to his other side. The weight of the vision made him feel claustrophobic. He got up and padded softly into the kitchen, put the water on to boil. He was pouring hot chocolate when he heard the front door open.

'And here we are!' the ghost said.

A woman giggled tentatively. 'Oh my. This is just *too* precious!'

'Come right on in, honey. Watch yourself, there—'

'Oops.' Something fragile crashed succinctly, with the *pock* of a ruptured light bulb. 'Oh, I'm so *sorry*. Here, let me—'

'Don't worry about it. All this junk belonged to my mother. I don't personally give a goddamn about crystal stemware.'

'Your *mother's*. I feel so terrible!'

'Shut up and come here a second.'

Then they were silent. Something warm breathed expansively through the house. Even in the kitchen Kenneth could feel the slow deep vital pulse, like that of a hibernal animal. Kenneth shut off the stove and ducked behind the refrigerator.

He heard footsteps in the hall. Someone's elbow cracked against the wall.

'Ow!'

'Careful, honey. This is the bedroom.'

'Oh my. Don't tell me you *sleep* on that? It looks like a museum piece.'

'So do you, baby.'

'I never had any idea the Ken Millar *I* worked with was such a silver-tongued devil.'

The switch snapped in the hall, the light went out in the kitchen. The hush resumed, the slow pulse quickened.

Is that really me? Kenneth wondered, alert behind the frost-free Maytag. Drunk, with a woman in the house? He heard a quick breath, a mutter. Something rustled.

A weightless sensation lifted in Kenneth's chest. The refrigerator clicked and hummed. He peered into the hallway, the steady shadows. They must be in the bedroom. He thought he discerned them, a dark, formless movement, protected even from refracted moonlight by the blackout curtains Mother had saved since the war.

Mutters, a small cry, staggered breathing.

Then, 'What's wrong?' A whisper, the woman's.

'Nothing. What are you stopping for? Give me a minute.' The urgent sounds were replaced by brisk, short whispers, like dialogue overheard at a clandestine business conference.

'What's the matter? Should I do something else?'

'No, that's not it.'

'What's the matter with you, then?'

'I don't know.'

'I'm sorry if I'm not exciting enough for you.'

'That's not it, honey. Don't be like that.'

In the kitchen the mug of cocoa slipped through Kenneth's fingers.

'What was that?'

'Nothing. Come back to bed.'

'Is somebody else in this house? Are you some kind of weirdo or something? Is that it?'

'Listen to me, Veronica—'

'I don't *feel* like listening. I'm getting out of here. All the men in this town are crazy!'

Fast footsteps in the hall. The chain rasped, the bolt snapped and the door slammed.

The ghost came into the dim hallway, turning his shirt-tail into his unbuckled trousers.

Kenneth emerged from behind the Maytag, sidestepped a Rorschach of cocoa and shattered ceramic chips and followed the ghost into the living room.

The ghost cinched his belt. 'So what are you looking at?'

'A real man,' Kenneth said.

'I had a little too much to drink. So what? It's her loss.'

'Whatever you say, Zorba.'

'Don't start in on me, pal. I sure as hell got a lot further than you ever did. At least *I* knew what I was looking for. Where you going?'

Kenneth opened the desk drawer, reached behind the walnut case. He removed a gun from the drawer.

'Oh, suddenly we're a real tough guy, huh? Not afraid of guns anymore, is that it?'

Kenneth lifted the gun with both hands, the way he had seen done on television programs. His teeth and hands clenched. The gun was soft in his hands, like clay.

'Christ, stupid. You have to release the safety first. Don't you remember the guy at the gun shop said—'

Kenneth pulled again. The gun made a feeble, hollow click, like a penny dropped into a pail.

'I really didn't think you had it in you. I never believed you'd pull the trigger. Good thing I hedge my bets.'

Kenneth lowered his unsteady hands. The gun slipped through his fingers and thudded on to the floor.

The ghost reached into his pocket. 'You'll need these.' He rattled something in his loose fist, stooped, placed the bullets on the carpet one at a time. 'One, two, three, four, five, six. There, now let's try it again. Ready, set? Go.'

Kenneth got down on his knees. His hands trembled wildly; his face felt swollen and hot. The objects loomed enormously on the carpet, as if viewed through a magnifying glass. He reached for a gleaming bullet. The bullet squirted out from between his fingers. He reached for the gun. The gun remained on the thin carpet.

'Now let's see how *you* like it,' the ghost said. 'Being stuck in the house all day without even a television to keep you company.'

Kenneth stood in the middle of the living room. He was warm in the white light of the overhead lamp. He stared at a tiny blemish on the wall behind the sofa, a tiny ridge of impasto which contradicted the otherwise smooth surface of paint. An entire world existed in that tiny blemish. Wild, forlorn, alien, crepuscular, a planet of rock and stone and cold, colorless sunlight. Kenneth knew he would be very happy there.

'I'm telling you, pal. Things couldn't be better.' The ghost sat on the sofa, drew a long breath from a Camel filterless, spat a sliver of tobacco. 'Ever since Veronica and I made up she likes me better than ever. It's not only my money, either.' He knocked an ash into the tall rock-cut crystal vase. Deep in the vase sparks smoldered among accumulated butts and black, twisted matches. 'She's really crazy about me. You hear me? Pal? Where are you?'

'Right here,' Kenneth said.

The ghost peered. 'Step out of the light so I can see you.'

A fine layer of gray ash settled over the black horsehair upholstery, Persian carpet and Renaissance revival side table.

A few months after the honeymoon Ken Millar sat in his brand-new Naugahyde Barcalounger. On one wall hung a crushed-velvet Elvis portrait, on another a seascape purchased from Starving Artists, Inc. The wall-to-wall shag carpet was described in the Ward's brochure as 'cream-of-pearl pile.' The television was on.

Veronica sat on the new paisley Hideabed sofa. 'This is nice,' she said absently, and turned the page of her *Vogue*.

Ken was watching *Today's FBI*. He had seen this one before. Ultimately the psychotic terrorist would be shot in the head as he ran for the helicopter.

'What do you think, Ken? Wouldn't it complement my eyes?' Veronica held up the magazine. The slick color page glared like an Indian signal from a hilltop.

'Beautiful. Now, if you don't mind, I'm trying to watch this.'

'I'm sorry if I disturbed you.'

Ken lit a cigarette.

'I sure wouldn't want to interrupt that damn television. It's the only thing in the world that seems to matter to you.'

'Why don't you go out and buy something. Just give me a break, will you?'

'Dr Silverstein says I spend money to compensate. Dr Silverstein says I need to get out more. He says he can tell I'm the sort of woman that loves to dance.'

'Screw Dr Silverstein.' This was the best part. The helicopter was beating down on to the parking lot of the Safeway, swirling old newspapers around. Mike Connors consulted his walkie-talkie.

'I might as well. It couldn't be any worse than what I've *been* getting.'

The psychotic emerged from the store. He gripped the gorgeous young prom queen by her long blond hair, gesturing severely with a Magnum.

'What was that?' Veronica asked.

'Sh!'

'Don't *shush* me! I asked what was that noise?'

Something rattled in the kitchen. The King William silver flatware.

'I told you before. These old houses have queer drafts.'

'These old houses are spooky, you mean. That settles it. We're moving to Van Nuys so I can be closer to my sister.'

The muzzle of the Magnum was wedged into the prom queen's ear. Her hair fluttered and tangled in the wind from the helicopter. The psychotic angrily shouted his demands over the noise of the rotors. If his demands weren't met, the girl was going to get it.

What if this isn't a rerun? Ken thought. Or what if the producers have substituted an alternative denouement, and in this version the psychotic escapes with his loot to Cuba, Argentina or even Paris, France? Perhaps this time it will be that oh-so-perfect prom queen's brain that litters the asphalt. Ken reached for his can of Bud.

'Damn,' he said.

'Oh, *really*, Ken. All over our new carpet. What's the matter with you lately? Can't you hold on to *anything*?'

DAZZLE

Dazzle was a dog with bushy red hair, fleas and an extraordinarily good attention span – especially for a dog. He was particularly fond of pastry, philosophies of language and Third World political theory. It was Dazzle's express opinion that unless somebody started paying the Third World a little concerted attention, serious consequences faced all mankind. Philosophies of language, on the other hand, were just a hobby, and when it came to pastry Dazzle preferred Sarah Lee Strawberry Cheesecake. There was more dog than dogness about Dazzle. Generally Dazzle knew how to keep his mouth shut, and strenuously avoided calling any attention to himself.

'The little doggy go woof,' said Jennifer Davenport, the youngest member of Dazzle's patron family, the Davenports. Jennifer was six years old. Whenever anybody visited they said how beautiful Jennifer was. Dazzle thought Jennifer was just okay. 'Woof, doggy. Be a good goddy – oops, I said doog goddy, I mean – ' Jennifer looked theatrically around at her family, who had positioned themselves conscientiously around the living-room television, but nobody looked back.

Doggies don't go woof, Dazzle thought, suffering Jennifer's cold hand on his nose. The *Canis familiaris* utters a guttural diphthong, much like the Mandarin Chinese diphthong, only less enunciated. Now why don't you leave me alone and go watch a little TV. Jennifer was already tempted. The television radiated warm noise and a flickering colorless haze that illuminated the faces of Father, Mother, Billy and Brad like nuclear isotopes. Mother was the Big One who fed Dazzle. Billy was the Little One who took him for the best walks.

Dogs don't like people, Dazzle thought. Dogs like dogs. Dazzle liked Homer, a resolute and well-groomed dalmatian who often roamed the park during Dazzle's afternoon walks, and Dingus, the hideous Lhasa apso who snorted at Dazzle through the slatted pine fence of Dazzle's backyard. 'Life's a game, Dingus,' Dazzle would say, contentedly pawing his rawhide bone and gazing up at the blue, translucent sky.

'Life's a game and you learn to play it by the rules, or else you learn to make everybody else play by your rules. You can either be the ruler or the ruled, and that's the crux, isn't it, old pal? That's the decision we've all got to make. Me, I'd rather live by the rules I'm dealt. I'm no high achiever, Dingus. I like my life. I eat well and get plenty of exercise, and I've pretty much got this whole damn yard to myself most days. Of course, I'm what you'd have to call an exceptional dog, but being exceptional's one of those things that takes a lot of effort, let me tell you. Being exceptional takes nothing but lots of hard work, work, work. Being exceptional just means lots of pain and suffering, pal, believe you me. Look at Kerouac. Look at Martin Luther King. They were exceptional, and where'd being exceptional get them? I'll tell you where it got them. It got them nowhere. It got them nowhere at all.'

Dingus snuffled against the pine fence. 'Dog smells,' he said. 'Food and water. Food and water and dog smells.' Dingus snorted and snuffled again, and eventually lay down in the warm dirt and fell asleep. In his dreams of quick rabbits, Dingus kicked. 'Rabbits,' he muttered in his sleep. 'Quick rabbits.'

Some days, though, Dazzle was so depressed he couldn't even get out of bed to go to the bathroom. He lay on his desultory, twisted blanket beside the water heater in the basement and awaited the occasional click of the thermostat and the rush of the gas fire which indicated Mother was washing dishes or doing laundry. Dazzle never knew what it was exactly. He just felt a sort of vague and indefinable anxiety, a certain fundamental sadness at the inconclusiveness of things. It was the way he felt when he saw a dead cat in the road. Dazzle hated cats, but when he saw them squashed and senseless in the spattered street he didn't hate them at all anymore. He sniffed at them; they didn't even smell like cat. They smelled like hot asphalt, transmission fluid and gasoline. Sometimes Dazzle just lay on his blanket for hours, contemplating the meaninglessness of dead cats. When the postman pushed mail through the grate he might try to emit a halfhearted growl, but usually he didn't bother. Eventually he would hear Billy's bike clattering on to the dirt driveway, and force himself shaggily to his feet, shaking off his loose, dandruffy hairs. Dazzle simply didn't have any clear idea what was bothering him, but did his best to keep up a good false front. He didn't want people to think he was just feeling sorry for himself.

'It's a good life,' he told Homer in the park. Billy was sitting on the softball bleachers with his friends and absently handling Dazzle's disenfranchised leash. 'Of course, it's a routine, we all know that. A dog's life, as they say. Dog days in a dog's life, and all that. But you take away routine and what have you got? Have you read the papers lately? Do you know what the world population figures are, going into the next century? Check it out, pal; check it out. And do you know what our President's official policy is on overpopulation? Well, hold still just a second and I'll tell you. Our President thinks any increase in population only creates a larger, wealthier consumer market. That's what our President thinks. The more overpopulated the world gets, the more Volvos we'll sell. The more canned spinach. The more Levi's. The world's going to hell in a handcart, Homer. Let me tell you. So routine, well, maybe it's a bit boring at times. But it's better than no routine at all – and you know what no routine at all means, don't you Homer? It means chaos, entropy, deindividuation, madness and death. How's that sound to you, Homer? Does that sound better than one square meal a day, and a warm blanket to sleep on? Does it, Homer? You tell me. Because maybe, just maybe *I've* had it wrong all these years.'

'Relax and eat a bone,' Homer said, imperiously panting, gazing off dreamily at a big black bird on a wire. 'Gnaw a bone and dig the yard.' Homer was a sage and sensible dog, Dazzle thought. But he was also, like every other dog Dazzle had ever met in his entire life, extraordinarily stupid. Even Dazzle's brave and obvious exercises in false bravado were lost on stupid dogs like Homer and Dingus. It was a very lonely world, Dazzle thought, this world of dogs.

Nights Dazzle suffered long knotty bouts of insomnia which arose in him as charged, eccentric monologues filled with delusions of grandeur and then, just as impossibly, plunged him into the depths of irony, self-mockery and suicidal despair. 'The best lives are simple lives,' Dazzle tried firmly to convince himself, unable to sit still even for a minute. He heard the mice in the garbage, the beetles on the walls. He got up and turned and turned again on his frazzled blanket. Then, of course, the irrepressible antitheses arose too. 'Simple lives are filled with loneliness, vacancy and self-deception.' For days he would go without eating, just gazing emptily at the liquefying Gravy Train in his big red bowl. Flies settled in it; then, at night, the mice came. It didn't really matter, he thought. Sometimes when he urinated on the newspapers, a few tiny drops of blood dripped out. His stomach twitched and growled; he experienced long energetic periods of flatulence. Some days he couldn't even bear the thought of facing the world's other dogs.

The family veterinarian, Hsiang the Merciless, prescribed antibiotics. This is life's real horror, Dazzle thought, prone on the ice-cold Formica table, closing his eyes and abiding the flea spray's aerosol hiss. Streams of black fleas spilled across the thin tissue sanitary sheet. We live and we die by the hundreds and thousands, Dazzle thought. And in order to live we visit our doctor. Dr Hsiang's gloved hands prodded and violated Dazzle in every conceivable way, and in many inconceivable ones. Dazzle shivered with terror, surrounded by the cold, antiseptic office, the menacing banks of glittering stainless-steel blades and instruments. This is the horror of life, he thought. This is life's trial, and then we die.

When Billy hid the antibiotic pills in tiny edible lumps of Dazzle's Alpo, Dazzle would carefully disengage and deposit them behind the hot water boiler where nobody, to his knowledge, ever cleaned. He had heard too much about the debilitating effect of antibiotics on the body's immune system, and anyway, he knew his grief was not merely physiochemical. It was philosophical, ethical and spiritual. It was a logical problem he would have to deal with. Intellectually, he knew he was on firm ground. Maybe life wasn't filled with all the excitement and challenge he might have desired, but, he reflected, you can't always change life. You can't always change history, Kenneth Burke once said, but you can change your attitude toward history. Dazzle had a bad attitude, for which he could only hold himself responsible. Fundamentally, Dazzle considered himself an existential humanist. This meant he didn't believe in God, but he did believe in guilt.

'Good dog. Nice dog. Dazzle is a nice dog,' the psychiatrist said, cradling Dazzle's freshly laundered blanket in his arms as if it were a baby. The psychiatrist was balding and slightly pock-marked; he wore thick wire-rimmed glasses. 'He looked a little ludicrous, if you want to know the truth,' Dazzle told Dingus later. The psychiatrist's name was Dr Bernstein, and Dr Bernstein told Mr Davenport that Dazzle suffered from acute feelings of insecurity initiated by a birth trauma and castration. ('I think it was castration,' Dazzle said. 'You want to talk about trauma, let's forget birth altogether. Let's talk about getting your balls chopped off. Bastards.' Dingus snuffled miserably.) 'Nice dog. Good dog,' Dr Bernstein said, burying his silly face in the blanket and sniffing audibly at it as if he, and not Dazzle, were the dog. Then he smiled. 'Dazzle smells nice. Dazzle's blanket smells nice.' Once each week Dazzle lay on the tiny hearth rug beside the electric fire, peering up at the visibly disturbed and often unsettling Dr Bernstein. Dr Bernstein pranced about, made barking and growling noises and offered Dazzle a red rubber chew toy which Dazzle merely contemplated lying there before him like a mantra or something by Wittgenstein. 'I think it's true what they say about psychiatrists,' Dazzle said later. 'They're all crazy. They're all fucking nutty as fruit bats.' As Dazzle said this, Dingus lay down in the dirt and began licking himself noisily.

Veterinarians, canine shrinks, other dogs, Big and Little Ones. Things seemed to be getting worse and worse rather than better and better, at least as far as Dazzle's state of mind was concerned. The Family even began to regard Dazzle with a sort of diffracted familiarity. 'Hi, Dazzle,' they said, without bending to pet him. 'How you doing, boy?' They looked genuinely concerned, but they also looked like they didn't really want to get too involved. Dazzle didn't know what to say. Every evening he watched them assemble around the glowing television and sometimes, out of the corners of their eyes, they watched him watching them. He sat and listened unemotionally to the news. The entire world was rapidly being transformed into a gigantic petrochemical dump, Dazzle thought. We are all being steadily infiltrated by carcinogens, toxins, radiation and some sort of irrepressible sadness which is probably the only underlying meaning anyway. Jennifer never snuck Dazzle into her room anymore so he could sleep on the big bed. By now, though, Dazzle had learned to prefer the garage.

*

Then one cold afternoon, while Dazzle was sitting in the backyard talking to Dingus, he noticed the gate was open. The latch had not engaged; and the wind was beating it gently against its creaking hinges. Dingus noticed it first, and began snuffling and darting back and forth against his own fence, insensibly sensing the sudden miracle of Dazzle's. 'Cats!' Dingus cried. 'Piss everywhere! Kill all the cats!' Dazzle watched his frenetic neighbor with a cool and cynical distemper. The bravest dogs in the world are the dogs who bark behind fences, Dazzle thought philosophically, and got up and went to the gate which, with a tiny pull from his paw, swung widely open and revealed the rolling hills of the Simon Hills tract estates covered with their uniform houses. With a sinking feeling in his chest, Dazzle stepped outside into the unfenced world.

For days and days he wandered aimlessly, urinating weakly on trees, lampposts and hydrants with a distracted, almost surreptitious expression, as if he were secretly determined to eliminate all the world's traces of other dogs. His legs carried him steadily and rhythmically in no direction at all, and for a while he preferred this sort of primal, nomadic state of disaffected consciousness. 'It's not the rhythm of the primitive we've lost,' Dazzle said out loud sadly, to nobody in particular. 'It's the rhythm of history itself.' Other dogs often appeared and sniffed at Dazzle, and with impatient formality, Dazzle sniffed back. 'Don't eat our food or piss on our posts,' the dogs said. 'Don't fuck our bitches.' Dazzle disregarded the blind caution of their first warning, contemplated the ironic sadness of their second. One night, while he tried to sleep in an alley, he was even approached by a bitch in heat. She was filthy and bone-hungry, and stank terribly. He looked up drowsily as she sniffed at him. 'Sorry, dear,' he said, watching her lumber off in her manic, erotic daze. We just bang around the world like that, Dazzle thought. We travel around the world banging into things.

Out here in the unfenced world, Dazzle's dream-life gathered strange energy and momentum. They were muted, vegetable dreams, filled with formless bodies and soundless words, and when Dazzle awoke he found himself inexplicably contemplating the migration of birds, the constellations of his youth. He remembered as a pup lying out on the backyard's green grass with his chew toys and identifying them: Orion, Taurus, Hydra, the Pleiades. His sharp canine eyes could discern the rings of Saturn, the moons of Jupiter. Space was filled with awesome distances and complications. It went on forever and ever. Quasars, pulsars, stars, galaxies, vast convoluted nebulas like memories, shattered planets and exploding stars. As Dazzle grew older and older all this universal wonder seemed to shrink and encapsulate him like a glove. He forgot his own wonderment, or at least considered it frivolous. 'I'm no adventurer,' he used to tell the equally youthful Dingus. 'I'm a house pet. I know how to keep my four paws planted right here on old terra firma.' Recalling now his own heartless and cynical affectations, Dazzle started to cry. He didn't understand this inexpressible sadness, and wished it would end and leave him in peace. He wanted to be happy in his yard again. He wanted his bland regular meals and his blanket. He wanted that hard, musty world in which he knew the locations of things. He desired the simple dreams before language began, and regretted his own smug complicity in the world's systematic disavowal of imagination.

Needless to say, however, Dazzle didn't find much imagination out in the unfenced world either. Instead he found rampant street crime, bulleting cars and buses, underfunded public schools, political corruption, sad songs, homeless families, bad meat and tall office buildings. 'Sometimes there's just nothing you can say,' Dazzle told himself, chewing morosely at a slice of stale bread he had pilfered from some addled pigeons. 'Sometimes there aren't any explanations. Or if there are explanations, they don't make you feel any better.' He slept in parks, alleyways, underneath parked automobiles, sensing his own true home diminishing in the world and reeling further and further away, like stars and nebulas and other planets in a universe of constant motion. All the stars in the world are hurtling further and further apart, Dazzle thought. Dissolution, heterogeneity and death. Even the Davenports were fading from the map of Dazzle's mind. Soon, Dazzle thought, the map will only contain direction and gravity and heat. It will lack a central landmark. There will just be the world, and me in it. It was hard for Dazzle to believe he could feel any more forlorn and helpless than he had when he was living with the Davenports, but now his sadness had become actually inarticulable. He couldn't even compose sentences about how he felt; he couldn't alleviate the weight of his misery with metaphors or figurative language of any sort. His monotonous legs carried him deeper into the world of noise and lights and cities. Sometimes he encountered stray coyotes, or even wild wolves who had gotten lost for years in the big cities and were now almost completely insane. They talked out loud to themselves and yelped pitifully at the least sudden sounds. They suffered from skin diseases, vitamin deficiencies and a wild, unexaggerated fear of all mankind. Howl at the moon, Dazzle told them. The moon's a bitch.

Dazzle was sleeping in an overturned trash can on a neglected and undeveloped Encino lot when he met Edwina. Edwina was a nice dog, though a little lean, who suffered from chronic indigestion and a severe overdependency on father figures. 'Feed me,' she said. 'Beat me. Hurt me. Love me.' She bowed her head as she approached Dazzle's trash can, sniffing suspiciously at Dazzle's unalpha-like yawn. 'Sit down and rest,' Dazzle said, shifting a little to one side. Edwina sniffed more intimately. 'Forget it,' Dazzle said. 'Go to sleep,' and Edwina did. This was the closest Dazzle had ever come in his life to a real relationship, and as such he blithely accepted much which, intellectually, he often considered impossible or irrational. For while most of the time Edwina was disturbingly submissive, at other times, entirely without warning or apparent provocation, she would take vicious and sudden bites out of Dazzle's rump or tail. 'Jesus Christ,' Dazzle would say. Edwina was like a beheaded or defaced street sign. There was very little understanding Edwina, which actually reassured Dazzle in a strange way, for Dazzle was a dog who for many years had felt he could understand just about anything, particularly dumb dogs.

Edwina didn't know anything, and relied on Dazzle to find food, evade dog catchers and traffic and some nights even to get to sleep. 'Everything's going to be okay,' Dazzle comforted her, gazing out at the starless, smoggy sky. 'You just relax and go to sleep.' Edwina was a hopeless case and even more screwed-up than Dazzle, and that was why Dazzle suspected, at times, that he just might love her. He paid her close attention. 'You're not going to eat that, are you?' he might ask, or 'Try taking a little bath in the pond in the park. You'll feel better.' Whenever Edwina was in heat, she would bring home the mangiest, most smelly and disreputable dogs she could find and fuck them in the bushes behind Dazzle's trash can. Afterward the spent and irritable dogs would try to pick fights with Dazzle, or bully him into giving away some of the food he kept wrapped up in newspapers inside his trash can. 'One of these days, Edwina, you're going to get fucked by the wrong sort of character,' Dazzle warned her after another very long night. 'You know what that means, don't you? Rabies. Yeah, you heard me. Frothing at the mouth. These enormous lesions develop all along your spine and inside your brain.

You'll be one fucking crazy bitch, Edwina. I mean, I'm just thinking about your health; I think I'm old enough to say quite frankly I've outgrown pride and all those other silly provocations to marital conflict. But I do wonder sometimes, Edwina, I really do. Where do you find these guys, anyway? I mean, do you actually go *looking* for lowlifes, or is it just that lowlifes are irresistibly attracted to you?' Sometimes Edwina made Dazzle feel extraordinarily weary and discomposed, and despondently he would root through his remembered past trying to unearth some forgotten scrap of nostalgia. He tried to formulate romantic images of himself back then, the lone roamer seeking truth in the world, learning about himself and his fellow dogs. But the romantic images rarely held together for more than a few moments at a time. When push came to shove, Dazzle had to admit it was much warmer sleeping in the trash can when Edwina was there.

Even though Dazzle had tried to explain contraception to Edwina about a million times, she never once paid him a moment's notice, and in May during her second month of term Dazzle decided they should head north, into the high and unpaved country where Edwina's litter could at least expect a few simple years of the reflective life before they were meaninglessly smashed beyond recognition by some errant and uncomprehending bus. 'Sometimes it's not the lived life that matters at all,' Dazzle told them. They were blind and sucking and crawling all over one another, still marked by bits of unlicked blood and placenta. Edwina lay deflated and insensible in the bole of a large tree which Dazzle had padded with leaves and a few scraps of charred blanket he had discovered near an abandoned campsite. They were living in Big Sur overlooking the rough Pacific, the gnarled and wind-shaped elms and shore. 'The lived life's just a big con too,' Dazzle told them. 'Events, possessions, sights, sounds, travel, achievement – oh, and what's the famous one – oh yeah, *experience*. It's all a big cultural snow job, if you ask me. It's primitive accumulation, the myth of the entrepreneur. There aren't any entrepreneurs anymore, kids. There's just ITT, Mobil and General Dynamics – and you know what they all thrive on, don't you? War, slaughtered and commodified animals like us, economic and political repression. Us unincorporated just have to do our best to carve out our own little alternative pockets of living. That's why the family's so important. I guess what I'm trying to say about all this nonsense is simply try and be happy with your life and don't worry too much about *experiencing* it. Let's all relax and enjoy ourselves a little. Let's all find a nice long pause together, okay, and not be in such a damn rush to get anywhere or do anything.' The squirming pups just squealed and sucked. Sometimes during those first few days Dazzle felt he was the one who had just been born.

If there was such a thing as happiness, Dazzle thought he had found it. The role of patriarch fitted him quite snugly, and he realized that even if he could not find any sort of subjective comfort himself he could at least meticulously envelop Edwina and her pups in comfort's illusion, that long, slow dream of culture Dazzle had always examined but never successfully comprehended before. It was Edwina's last litter, and since she didn't desire to get fucked that much anymore they were able to construct a relatively stable family environment. The pups grew with sudden and frightening alacrity, and there always seemed to be one or two of them pulling at Dazzle's tail or pawing at Dazzle's face. Dazzle hardly slept at all anymore, and generally he preferred this dulled, unquenched fuzziness of brain and perception. It's best to keep the old brain a little blurred, a little battered, Dazzle decided. He had cleared out a small cave underneath an outcrop of black igneous rock on a mountainside. As the pups grew, he trained them to maintain a system of revolving security watches around their home, and drilled them in defensive techniques and maneuvers.

'A man?' he asked them.

'Hide,' they said.

'Wolf?'

'Submit.'

'Bear?'

'Run.'

'Inexpressible sadness?'

'Run.'

'Restless, unhappy dreams?'

'Dream again.'

Often in the middle of Dazzle's patient drills, while the addled and hyper pups were growing distracted by buzzing flies and high birds, Edwina snuck up behind Dazzle and took a quick, nasty bite out of his ass.

'Jesus Christ,' Dazzle said.

*

There was a smooth diurnal rhythm to life now, Dazzle thought. You could feel the safe beat of the entire world in your blood, your heart, your dreams. Half-asleep at the mouth of their cave, he liked to listen to Edwina and the pups snoring and contemplate the stars outside again. Pisces, Cassiopeia, Ursa Major and of course the craters and mountains of a vast and irreproachable moon. This is where the cycle ends, he thought, if a cycle it is. It's that convergence of the stars and the blood, the moon and the heart. It's not the world of men. It's not the Davenports' smelly garage. It's not Dingus urinating on everything. It's not clinical depression, or obsessive, convoluted thinking. It's not even barking at the mailman.

There were still days when Dazzle would slip off into the nearest town and check out the newspapers. Islamic fundamentalism, AIDS, the international debt crisis, yuppie liberals, adamant right-wing perjurers. It's not to disavow the world that I've left it, Dazzle thought, and made his mark on the *Examiner*'s Op Ed page. It's to live in the world I've always before disavowed. If he moved quickly, he could pull a nice steak from the grocery's refrigerated cabinet and sneak quietly out the back door like some innocuous delivery boy.

Periodically, though, Edwina grew ill and somewhat disaffected, lying alone for days at a time gazing insensibly at the blue sky beyond their tidy and self-sufficient cave. 'Melancholy,' Dazzle wondered. 'Sad reflections. Lost love. Dead friends.' But Edwina never told him what was on her mind; she just growled distantly at him. She never even bit him anymore, and eventually Dazzle realized she was suffering from physical rather than merely philosophical distress. The whites of her eyes grew sallow and bloodshot. Her breath was bad, and she suffered frequent discharges of diarrhea. Small rashes formed occasionally on her back and stomach, and eventually Dazzle diagnosed a low-grade infection, perhaps septicemia, or a common form of acute gastroenteritis. Dazzle recalled the library of antibiotics he had so smugly discarded behind the water heater in the Davenports' garage. You can't go back and change some things, he thought. He liked that world better, the simple one of medicine.

Early on a Monday morning Dazzle descended to the town with Flaubert, the laconic and reserved pup who, like his brothers and sisters, was really a pup no longer. Flaubert was developing assurance and a quick stride. There was something wild about Flaubert which Dazzle didn't understand, something which Flaubert had either inherited from his mother or his uncivil upbringing. It wasn't just his eyes, for he carried a certain alertness in the very poise of his musculature. 'The world's crisis is a crisis of representation,' Dazzle explained as they descended the mountain. 'We're always representing our lives one way or another. We never live them. We never even live them *as* representations, which is an idea I've been giving a lot of thought to recently.'

Alpine was a minimal town which contained a small grocery, a pharmacist's, an abandoned movie theater, a Woolworth's, which had been recently converted into a Bill's Jumbo Discount House, and approximately six hundred people. 'There's a hidden continuity between signs and things, thoughts and world. Our fears of discontinuity are a fiction, actually, but one which we must be maintaining for some reason. Our anxieties about the world, things, other people, that world which doesn't conform to our dreams of it. We're letting those anxieties determine our world. Instead we should try to determine the world for ourselves.' Coolly, Flaubert loped along like a wolf; he didn't say anything. Dazzle thought Flaubert was starting to look a little bit like Warren Oates in *The Wild Bunch*. 'They're anxieties because we can't admit the validity of our own dreams,' Dazzle said aimlessly. 'That's what the world keeps telling us, you see, and that's what makes us so goddamn miserable. We believe what we're told, even when we're told to believe in everything but ourselves. I'm not trying to sound like some adolescent solipsist or anything, Flaubert. I'm not saying we should deny the world or anything. I'm just saying, let's give our dreams half a chance too. Let's maintain some faith not only in the world but in our dreams of it.'

They had come to a stop across the street from the Mercury Pharmacy, where the pharmacist, a tall man named Bill who wore a white jacket and patent leather shoes, was outside on the front curb training his guard dog, a large mean-looking Doberman whom the pharmacist referred to as Dutch, but who referred to himself in his most secret thoughts as Jasmine. The pharmacist pulled sternly at the Doberman's gleaming stainless-steel choke collar; at the same time he showed the captive dog a handful of chicken biscuits. 'Sit,' the pharmacist demanded impatiently, and gave the collar another sudden pull. 'Sit, Dutch.'

'Chicken biscuit,' the Doberman said. 'Chicken biscuit biscuit.'

'Sit. Sit *down*, Dutch. *Sit!*' the pharmacist said.

'Maybe he doesn't want to sit,' Dazzle said out loud, but nobody in the world was listening. 'Maybe he just wants his goddamn chicken biscuit. Maybe he just wants to eat his goddamn chicken biscuit and then take a nice long nap.'

When Dazzle was just a puppy his favorite television program had been called *Lassie* and had starred an attractive Scottish collie of the same name who saved members of the human family she lived with each week from various life-threatening situations. Lassie dived into raging rivers and burning buildings. She stood up against wild bears and men with guns. Lassie was a brave dog, Dazzle had thought, but an exceptionally foolhardy dog as well. 'Save yourself,' Dazzle would cry weakly, whimpering a little under his breath at the terrible trials and misfortunes endured by brave dogs everywhere. 'Run like hell. Timmy can take care of his damn self.'

'Sit,' the pharmacist said. It was a warm day, with only a few high white clouds. 'Sit *down*.'

The less and less I understand, the simpler everything seems, Dazzle thought, and at his signal, Flaubert took off and broke the pharmacist's grip on the Doberman's collar like a sprinter breathlessly striking the victory ribbon.

'Cats!' Flaubert cried, dashing off down the street. 'Cats!' The Doberman, with a brief flickering expression like the lens of a camera, poised and then, with a sudden start, took off after Flaubert. The pharmacist took off after him.

'Sit!' the pharmacist shouted, running and shaking his gleaming choke collar at the bright sky. 'Heel! Stop! Sit!'

Without a moment's hesitation, Dazzle loped into the pharmacist's office, found the Prescription Out tray, and snapped up one hundred capsules of 250 mg tetracycline and fifty 100 mg erythromycin. Then, with a flourish, he ascended again into the high mountains.

WHITE LAMP

Every morning Aunt Doris brought hot Cup-a-Soup to her great-nephew Cecil in a tall Styrofoam cup. Whenever Cecil visited Southern California he slept in his battered red Toyota in the parking lot of Aunt Doris's retirement hotel, and was usually still asleep when Aunt Doris came down to wake him. Sometimes his engine was running, his car heater turned up high; an easy-listening station was usually playing on the scratchy radio, and his entire car smelled of stale cigarettes. 'Here's your soup,' Aunt Doris told him. 'You need a cup of hot soup to warm you up.'

When he opened his eyes, they were red and blurry. 'What's that?' he asked. He reached automatically for his cigarettes from the dash.

'It's hot soup,' Aunt Doris told him. 'It's hot soup to warm you up.'

Aunt Doris felt guilty she couldn't let Cecil sleep on the floor of her room, but her room was very small, and the epidermis situation was a problem simply too dangerous to contemplate. 'First your epidermis gets into the rug and then my cats pick it up,' Aunt Doris told him. 'Then I pet my cats and it gets all over me. Before you know it, your epidermis is everywhere. It's in the food and in the water, it's in the towels and in the dishes. Did you know that your average American loses six to eight pounds off his epidermis every month? This is a fact. This is a fact which I have very carefully checked into.'

Aunt Doris lived in room 224 of the Hotel San Clemente, which advertised itself in its brochures as 'retirement living by the sea.' Her room contained a clock radio with a broken hour dial, a single bed, a hand-held rechargeable vacuum and a copy of Mary Baker Eddy's *Science and Health with Key to the Scriptures*.

Health was one of Aunt Doris's primary concerns, and she tried to maintain her proper vitamin intake, keep an optimistic attitude about life and exercise regularly by taking the hotel's carpeted stairway rather than its grinding, inconstant elevator. After delivering Cecil's soup she would fix herself a cup of hot Nestlé's Quik and a Carnation Breakfast Square, which together provided more than twice her daily vitamin and iron requirements. In fact, if Aunt Doris were to list her primary concerns on a sheet of white paper (which she was often fond of doing), her list would just about always include health, grooming, rent and utilities. Generally she tended to include 'grooming' out of a vague sense of decorum, since she rarely showered, and even then only to swab her body gently with a damp sponge. Water, she believed, drained the body of its fundamental energies and sustenance. Good health and long life required that people industriously preserve the integrity of their own skin.

Every day, however, Aunt Doris made sure to brush her teeth with mineral water and Crest. She still had a perfect set of white teeth, and in eighty-three years had yet to suffer a single cavity. Every month she received Social Security, Medicaid (for her back) and a generous pension check from Jack's Hardware, the former employer of her deceased husband, Darrell Dwayne Mulhall. Every Christmas Jack's Hardware sent her a nice card as well, personally signed by Jack's surviving son, 'Pittsie,' who, as Aunt Doris dimly recalled, had seemed like a very nice man the afternoon she met him at the racetrack with Darrell Dwayne Mulhall. Sometimes Aunt Doris added that to her list of primary concerns: the past. The past was a lot like the other things on the list, really. Like health, rent and utilities, the past posed certain standard and often unacknowledged obligations. The past had determined Aunt Doris's name, personality and place of residence; the past was not a force so much as a sort of obstinate weight. It never really motivated you toward anything, but merely accumulated like battered cardboard boxes filled with disused but invaluable mementos. Sometimes Aunt Doris felt she would be much happier as an amnesiac, like one of those confused and often beautifully tragic characters on daytime television melodrama; other times she thought she would be much happier without any past at all.

She rarely slept at all anymore. After Ted Koppel, whom she felt herself distantly attracted to, she might watch a movie, or a rerun of one of last Sunday's community service programs. Sometimes she listened to all-talk radio, or read the newspaper, but usually she just sat by the front window in her tiny room and gazed down at central San Clemente, where alternating streetlights presided over the empty, adobe-tessellated streets, humming and clicking like contemplative machines. One night she saw a young man coming up the street, carrying a large television. He was hurrying, and disappeared into the alleyway beneath her window. In the dark and from a distance, he reminded her of her great-nephew Cecil, and she felt a brief pang of remorse. Cecil was Dotty's grandson, though Dotty herself had never actually lived to see him. Dotty had died when she was only nineteen, shortly after the birth of her first child, Agnes, who was Cecil's mother. Whenever Aunt Doris thought of her dead sister Dotty, she hoped Cecil was sleeping safely and comfortably in his untidy car. Tomorrow she would fix him toast, too. Cup-a-Soup and toasted Wonder bread. If she remembered.

Sometimes Aunt Doris couldn't tell if she was dreaming or not. Throughout the night she could hear definite footsteps outside in the corridor, and sometimes she would get up from the bed where she lay sleepless, pull on her robe and go to the door and open it. It might be Carl from across the hall, fumbling with his massive knot of keys. Or it might be Ted, pacing the corridors with his aluminum walker. But sometimes it was people Aunt Doris had known and loved during her life, many of whom were now deceased. Sometimes Aunt Doris saw her father in the corridor, or her mother. She saw Dotty, or her deceased husband, Darrell Dwayne Mulhall. Sometimes she saw familiar but unidentifiable children, who gazed up at her with unspeakable motives. The people from Aunt Doris's past never said anything to her; in fact, Aunt Doris perceived them as if through a thin veil of heat. There was something heavy and oppressive about the corridor's atmosphere at these times, and Aunt Doris would quickly shut the door, afraid it might infiltrate her room. Then she would lie back down on her bed for a while, wondering if she was still dreaming. But it didn't seem like dreaming, because she never felt herself fall asleep or start awake again. It was as if dreaming and life had become somehow continuous, like the past and the present, like youth and old age. Aunt Doris knew she wasn't young anymore. She knew she was very old.

*

One evening when Cecil was still visiting, Aunt Doris opened her front door and saw Emmett Stanley standing in the hallway. Emmett was carrying a large black plastic flashlight and looking very young for his age. He didn't recognize her at first, though Aunt Doris recognized him right away.

'Sure,' he said finally. 'Doris. Of course. Doris Kelly.' Doris Kelly had been Aunt Doris's maiden name when she was growing up in Fullerton.

Emmett Stanley was the first man who had ever made love to Aunt Doris, and now, by sheer coincidence, he lived in Aunt Doris's own retirement hotel. 'I'm looking for the water heater on this floor,' Emmett said. He was wearing very baggy coveralls and a torn long-sleeved thermal shirt. 'Do you know where the water heater for this floor is?' He swung the flashlight beam up and down the corridor.

Aunt Doris showed him the door to the water heater, which was just two doors down from her own. The lock was broken, so Emmett didn't have to look for his keys.

'You never came back, Emmett,' Aunt Doris said. 'You told me you'd come back for me, but you never came back.'

Emmett leaned into the cramped and musty closet. The flashlight projected the water tank's interior across the corridor walls, and Aunt Doris found herself standing in the enormous shadows of spigots and winding copper tubes. 'It looks like the burner's gone,' Emmett said. 'I'll have to drive into El Toro and get a new burner for it tomorrow.'

'Maybe it's because we didn't have a real relationship,' Aunt Doris said. Emmett was pushing shut the closet door; it dragged roughly against the carpet. That afternoon at Baskin and Robbins, Aunt Doris had overheard one young man telling another, 'Yeah, I'm seeing someone. But it isn't really a relationship.'

Emmett Stanley lived in a minimal storeroom in the hotel basement. The door to his room was located behind the matching Maytag washer-dryers. To get to it Aunt Doris had to step across heavy black tubes, coils of ominous red and brown wiring and large, birthlike clumps of blue lint. She made sure she was careful of her back. The storeroom was very dark as you approached and pulled open the perforated plywood door. 'I must confess,' Aunt Doris said. 'I do think about you sometimes, Emmett.' She stepped down on to the cold concrete floor. She felt slightly dizzy for a moment as the dark room grew gradually more luminous and coherent around her. The air was heavy with thick, whirling motes.

'That's nice,' Emmett said. He was suddenly beside her, and slamming his steel toolbox on to the floor. 'Would you like some coffee or some tea?'

Emmett's room underneath the hotel contained two tattered stuffed chairs, a large mattress on the floor, a couple of throw rugs and a stained wooden end table between the chairs, which held a large glass ashtray and a half-finished cigar. One of the stuffed chairs contained Emmett's wife, who was reading the *Times Sunday Home Supplement.* Behind her chair, balanced atop a ledge of concrete, sat a white lamp with a beige shade. The base of the lamp was chipped and dusty; the bulb itself emitted a dusty, nebulous sort of light. In some ways it didn't seem like light at all, really. It was rather like some chemical or nuclear discharge that actually altered the deep cellular matter of things rather than merely illuminated them. Mrs Stanley, however, seemed to have no trouble reading underneath it; but then, she was also wearing glasses.

'Hi, honey,' Emmett said.

His wife looked up. 'What?' she shouted. 'What?' She was the most extraordinarily old woman Aunt Doris had ever seen, with glowing blue veins in her hands and a wide, translucent forehead. Aunt Doris thought the old woman might collapse at any moment; then only the raw, humanless stuff would remain, like the clumps of lint behind the dryer.

Emmett looked at Aunt Doris again. 'Did you say you wanted coffee, or what?'

Emmett took his cigar from the tray and lit it with a burnished steel lighter. It was, Aunt Doris learned later, a retirement gift Emmett had received from the Torrance Retread and Rubber Company nineteen years ago. That would have been 1967, the year Aunt Doris's husband, Darrell Dwayne Mulhall, had died.

Every morning Aunt Doris brought Cecil his Cup-a-Soup and dimly attempted to converse with him. Cecil referred to his job as something that included a lot of 'trading and bargaining,' and usually the backseat of his car was filled with various surplus goods he had acquired during his unspecific 'negotiations.'

'That's a nice set of golf clubs,' Aunt Doris might say.

'Yeah, I guess,' Cecil said without turning to look, gazing dazedly into the Styrofoam's hot noodles and viscid yellow water. His T-shirt was torn, and he scratched himself.

'Where do you sell nice stuff like you get? Where would you sell that nice electric typewriter, just for example?'

'Places I know.'

'Where would you sell that nice toaster oven? I'm not trying to be nosy, but I could use a nice toaster oven myself. How much might you ask for that?'

Cecil scratched at his stubbly face. The scratching made a loud noise, like static on a radio. 'Five dollars, I guess,' he said. 'Actually, make that ten.'

Aunt Doris took the toaster oven up to her room and gave Cecil ten dollars the next morning with his soup. Now she could fix toast and hot sandwiches for her friends whenever they came to visit. Somehow Aunt Doris felt less empty than before. She gazed out the window with a less aimless abstraction. She even noticed things sometimes. Boys on bicycles, young girls in summer dresses. She tried to remember these things so she would have something to talk about with her friends, Mr and Mrs Emmett Stanley. 'I saw a girl with beautiful blond hair today,' she would tell them. 'She had beautiful blond hair all the way down past her backside. When I was little I had long blond hair too, but then one day my mother chopped it all off.' Life seemed very peaceful and safe for the first time that Aunt Doris could remember. For many years Aunt Doris had thought that every person she had ever loved in her entire life was dead, so it came as something of a relief to find Emmett again, even if he was married. Aunt Doris looked forward to the late evenings when Emmett would be finished with his handyman chores and she could go visit him.

'Do you feel good about your life?' she might ask him. He sat in his stuffed chair smoking, and Aunt Doris would lower herself awkwardly onto a large stuffed pillow, her veiny legs splayed out in front of her like awkward debts. 'I mean, do you think life has been good to you?'

'How do you mean?' Emmett was peeking over his wife's shoulder. Mrs Stanley always seemed to be reading the same Sunday supplement.

'I mean, have you ever wished you could live your life over again? Have you ever wished you were young, and knew all the right decisions to make? I mean, it doesn't seem like being happy would ever be such a big problem, but frankly I was never happy at all, and my life was just one big mess actually. I mean, after you went to Oregon, I married a man named Darrell Dwayne Mulhall, not because I loved him but because my parents told me it was a good idea, and since I was pretty stupid when I was young, I completely believed them. Darrell Dwayne Mulhall was the most horrible man the world has ever known, which is not a nice way to speak about the dead, I guess, so maybe I should stop.'

Usually Emmett would fall asleep in his stuffed chair, the cigar smoldering beside him. Sometimes Aunt Doris would crush out the cigar for him and douse it with a little water from the spigot. Sometimes she covered him with one of the blankets from the bed. She never thought to check if Mrs Stanley was asleep or not. All she ever checked was to make sure that the white lamp was still on when she left. Without anybody having to tell her, Aunt Doris knew that the white lamp should never be turned off, even for a moment. The storeroom was so impossibly dark that if you turned off the lamp you might never find your way in or out again.

'You must get this cozy room for free since you're the hotel handyman, don't you, Emmett?' she once asked Emmett Stanley.

'Actually, they just give me a good discount,' he said. 'Only two hundred and fifty dollars per month. But that includes utilities.'

Cecil said, 'It's a Gillette Foamy hot foam dispenser. It's still unopened in its original packing case. There's still an original warranty on this sucker. You buy something like this in the store it's gonna cost you thirty bucks easy. I'm talking ten, twelve bucks – and that's not including a nickel for my trouble. Okay, make it ten bucks. You can give it to your boyfriend for Christmas or his birthday or something. I'm telling you. I can get you lots of good deals on stuff like this.' Cecil placed the box down firmly on the countertop beside his coffee. He was sitting in the Travel Inn, which was located immediately next door to the San Clemente Hotel. The large clock on the wall said nine forty-five.

'Did you want to order something or are you just going to sit there yakking all night?' Debbie slung the billing pad against her hip pocket; her left hand scratched aimlessly at a starchy gravy stain on her left shoulder. 'You sure like to work your mouth,' she said.

Cecil gazed speculatively into Debbie's cleavage, as if he might have spilled some change down there. 'I guess I'll just have some more coffee.' Cecil lit a cigarette. He rested his left hand on the Gillette Foamy box. The weight and density of the box felt reassuring, like gravity.

People just don't know what's good for them, Cecil thought, three Manhattans later at Goody's bar down the street. Goody's was filled with crew-cut boy Marines, obese women with bad complexions, and interchangeable men in their late fifties with sunburned, wrinkled faces, tattoos and bushy sideburns. If you don't take advantage when opportunity knocks, boy, don't expect opportunity to just keep coming around. Opportunity's not something that comes to you. No sirree, boy. Opportunity's something you take. Otherwise it's gone for good and you might as well retire. You might as well cash in your chips and leave the game to us professionals.

At that moment Cecil looked up into the wide mirror behind the bar and saw Debbie enter through the swinging front door. She was still wearing her waitress uniform. She carried a lit Marlboro in her poised right hand as if it were a VIP pass.

'Hey, beautiful,' Cecil said. 'Off work and time to play, huh? Why don't you let me buy you a drink?'

'Get lost, creep,' Debbie said, and went into the large, smoky adjoining room where her boyfriend was playing eight ball.

It made little or no sense to Cecil. You ate, slept, took impossible mortgages, grew old, died. Maybe if you were lucky you'd live till you were a hundred, sleeping with your mangy cats and pacing restlessly around the same damn hotel every night, up and down the same stairs night after night, hour after hour, up and down. By the time Cecil returned to the parking lot of Aunt Doris's hotel it was almost two a.m. Across the building's white adobe face many of the grilled windows were still brightly lit, curtains drawn back to reveal indistinguishable old men and women pacing about their rooms, myopically peering down at the parking lot as if even at this late hour they expected company. The place actually stank of old people, Cecil thought. Old people smelled like junk food, pharmacies, stray clothes in a laundromat. He could hear the distant rush of the sea. I'll be damned if this is my last stop, Cecil thought, and forced the back door lock with his Shell card. The Shell card was embossed with the name of its original owner, someone named Harold T. Willoughby. The wide, half-lit foyer was filled with hideous velour furniture. The pale adobe ceiling was water-stained, and someone had overturned a tall plastic and chrome ashtray beside the manager's door, spilling ashes and butts and crumpled gum wrappers everywhere.

The door to the basement was open, and Cecil switched on the light. The lacquered yellow walls were suddenly very bright, and he descended into the laundry room and discovered the large twin Maytags. The door to the storeroom hung loosely open; Cecil turned on his pocket flashlight, swinging the beam around the low, musty room. 'Bull's-eye,' he said, and took the Sony cassette player, the hand-held rechargeable vacuum, the dusty box of tools and, after a moment's hesitation, the chipped ceramic white lamp, which he could use once he got a new place. Then he decided to drive to Lompoc, where a friend had offered him the accommodation of his living-room sofa.

The fog that night along Highway 5 was heavy and enveloping and close. There were times when Cecil could hardly discern the front end of his own car. But even then, when the actual road seemed to disappear and the fog wrapped you up in its own world of the invisible, you still felt as if you were surrounded by a peculiar sort of meaning. Around four a.m. Cecil pulled over to the side of the road and fell asleep with the engine running and the heater blowing. He didn't dream at all that night.

There were so many questions Aunt Doris still wanted to ask. Did Emmett believe in God, for instance, or predestination? Had Emmett supported the war in Vietnam? Did he have any children or grandchildren? Aunt Doris had always wanted grandchildren, but she couldn't have babies, or else Darrell Dwayne Mulhall couldn't have babies. (She and Darrell Dwayne Mulhall never really did figure that one out.)

Then one night Aunt Doris descended into the laundry room and discovered only darkness. She switched on the overhead light, which was pale and unsteady. The perforated plywood door was bolted shut with a large brass Yale lock. 'Emmett,' Aunt Doris said. 'Emmett, are you home? Mrs Stanley? This is your friend Aunt Doris from upstairs. Are you sleeping? Would you like me to come back later?' Aunt Doris stood silently in the empty laundry room for a long time. The darkness was dense and bristling behind the plyboard door. Cloudy, complicit, brushing against things. Or perhaps that was just the motes of dust, or the dead remembered rays of the white lamp.

Whenever she thought about Emmett Stanley she remembered the white light from the lamp. In fact, when she tried to recall any particularities about Emmett himself she drew a complete blank. She couldn't recall his face or the texture of his voice. She couldn't even recall the first time they made love. It wasn't the sort of event one was supposed to forget, and sometimes Aunt Doris even doubted her own abstract certainty that anything had happened at all. At times like these she lay on her bed and listened to Daisy pawing the cat litter. Then Daisy jumped on to the windowsill and sat there motionlessly for hours, like a white ceramic sculpture of a cat. Aunt Doris tried to recall Emmett Stanley's face, but all she could see was the hard, pulseless white light, not really light at all so much as a sort of atmosphere, something breathed rather than seen. 'Sometimes I think I might like to live my whole life over again,' Aunt Doris said out loud, to nobody in particular. 'Here I am, an old woman already, and I don't ever remember being happy in my entire life. I lived my whole life with Darrell Dwayne Mulhall, and then when Darrell Dwayne Mulhall died I really didn't have that many options left open to me. There aren't a whole lot of options left open to you when you're sixty-four.' Aunt Doris looked around her empty room. Daisy had spattered brown cat food all around her yellow bowl. Tiny brown cat prints tracked the beige tile floor and disappeared into the thinning carpet like a primitive mandala.

After a few weeks Aunt Doris stopped descending to the laundry room at night and decided Emmett Stanley and his wife had simply moved without telling her. Perhaps one of them had died, and the survivor was too stricken with grief to contact other living people. Or perhaps Emmett and his wife had made a suicide pact many years ago, so that if one of them died the other would take his or her own life at the first opportunity. Aunt Doris thought that a suicide pact was a very romantic idea, and firmly decided that if she were ever to fall in love with someone deeply and truly they would have a suicide pact together too.

Time began to radiate rather than flow. Events didn't occur in the world around Aunt Doris so much as strategically assemble, like places on a map around some large city or natural resource. Doors opening and closing up and down the hall. Voices in the street at night. Marines with blaring ghetto blasters descending the long, winding pavement to the beach. Daisy forever digging in her cat box as Aunt Doris lay on her back on the cold bed, listening to the world outside and watching the bright, angular sunlight sliding across her floor. Aunt Doris saw the open can of kidney beans on the sink, but she didn't feel like eating kidney beans. The refrigerator door was open, but the bulb had gone out, so the food remained cold but unlit. She could smell the baking soda and moldering cabbage, the milk as it turned rancid, the sour cheese and fruit. And Daisy pawing at everything, and knocking down the Kitten Chow from the high countertop. It sounded like gravel spraying, and then the gravel shaking in the cat box and the dense odor of Daisy's excrement. It wasn't Aunt Doris's life at all; she finally understood that now. It had always been Darrell Dwayne Mulhall's life, Darrell Dwayne Mulhall's desires, Darrell Dwayne Mulhall's house and car and job and pension. Now that she thought about it, Aunt Doris pretty much hated everything about Darrell Dwayne Mulhall. He was a hard object in the room, while Aunt Doris was just drifting, drifting. Darrell Dwayne Mulhall didn't make a move or a sound. He just sat there with his newspaper, scratching himself and farting all the time, while Emmett Stanley made love with some other young woman on a rattly cot in the back of his parents' trailer. That other young woman hadn't been thinking about anything while it happened. That other young woman had only wondered why she had been afraid, why the cot rattled so loudly, why they bothered. And then Emmett Stanley went away, into the white light, into the long, pulseless drone of history, the history of other people Aunt Doris would never know.

Every day Aunt Doris lay on her cot and waited for the mail to arrive. Lying there with her hand across her heart, Aunt Doris could hear the mailman on the porch, in the foyer, at the boxes. Maybe there was a letter. The mailman's keys clattered against the boxes. The mailboxes closed shut again. The white light was accumulating in the basement storeroom just like electricity, just like time. The white lamp's light was moving up the cement staircase, into the elevator shaft, suffusing the pale walls and ceilings of the foyer, infiltrating everything, every room and every body in the Hotel San Clemente. The white lamp was still on, Aunt Doris realized, and this meant that Emmett Stanley and his wife hadn't moved or died, but must have just returned from some brief vacation. The white lamp was on, and they were reading and repairing irreparable appliances down in their minimal home behind the washing machines. The white lamp was on, and there was a letter in Aunt Doris's box. Aunt Doris closed her eyes. The letter in her mailbox said:

Dear Aunt Doris,
I know it's a shame not to have written you in so many years and now only because I have bad news but that's the way we are I guess, we've never been much of a letter writing family I guess. But your nephew Cecil was arrested last week on Highway 5 for possession of stolen property, and Alex and I are worried that Cecil may have repaid all your kind attentions and accommodations by taking valuable items from you or your friends there at your nice hotel. We hope maybe he hasn't, but here is a list which was provided to us by the Buellton Police Department, whose address is enclosed in case you must contact them in any way. Maybe Cecil is a very confused young man and we hope he will grow out of his bad deeds, but sometimes it's hard to be very positive thinking about somebody like Cecil who we have loved and tried to teach true Christian values to time and time again. Maybe you could visit us this Christmas, and we will send you a bus ticket. It would be nice to see you again, and your new niece Daphne (who isn't new at all really since she is already eleven years old – time sure does fly!) would like to meet you. Here is the list as you will find printed below:

1 Sony portable b/w TV
1 Sony cassette player
2 Sw. army knives
1 Sears cordless phone
1 Canon prt. typer
1 partial Ency. Brit.
1 clarinet
3 cases Pepsi
1 white lamp
1 Sm Cor. word proc.
1 pocket calculator
1 power drill

Well anyway I hope that none of these things really are yours which would make both Alex and I and I am sure Cecil too feel very bad indeed.
Your loving and very apologetic niece,
Agnes

In a minute, Aunt Doris would get out of bed and go down to the mailbox for her letter. In a minute. But first she would close her eyes. She would rest her eyes for just one minute, and then she would go downstairs to the mailbox for her letter.

HEY HEY HEY

The same afternoon her boyfriend Charlie tried to stab her to death with a plastic fork, Sarah Stanford moved her nine-year-old son and principal possessions to her mother's house in south San Francisco. 'Charlie says he wasn't trying to stab me,' Sarah told her mother. 'Charlie says he was just *ges*turing with it. I'm sure people *ges*ture with forks the way Charlie gestures with forks. You could rob a gas station gesturing with a fork like that. That's what I told the police, anyway.'

Sarah's mother was sitting in the navy-blue Barcalounger with her well-thumbed copy of the *Bhagavad Gita* while Sarah's son, Raymond, was lying stretched out on the sofa with his *X-Men* comic book. Like his grandmother, Raymond wore a pair of wire-rim bifocals.

'The police just looked at me like I was some sort of nutcase. Of *course*,' Sarah declared affirmatively, as if she had just discovered uranium, 'the police were both men. The people that run the police department – they're all men, too. Charlie's a man – at least on his better days he is – and I'm a woman. So who do you *think* they're going to believe? I'll give you two guesses. Which means that now Charlie's sitting there free as a bird in *my* apartment, eating *my* food that I cooked. And here I am with a nine-year-old child on my hands trying to start life all over again.'

'Sometimes I worry about you, Sarah,' Raymond told her later that night after Sarah's mother went to bed. Sarah was tucking stale, moth-nibbled sheets into the sofa's corduroy cushions. 'Why don't you try calling Charlie and talking things over with him for just two minutes? We can't stay with Grandma forever, you know. I mean, the longer you keep making a big fuss about everything, the harder it's going to be on your pride in the long run.'

Sarah stuffed the side cushions into a large pillowcase emblazoned with a six-gun toting Yosemite Sam. Then she swung the spare blankets at Raymond and knocked him on to the sofa. She showed him the rectangular Band-aid affixed to the back of her wrist, as if impressing him with the late hour on a watch. 'You know what that is, Raymond? I'll tell you what that is. That is where your daddy tried to stab your mommy to death with a kitchen fork. Now why don't you go to sleep and give me about eight hours peace and quiet.'

Raymond sat clutching the blankets to his thin chest, blinking myopically, his glasses on the bureau. Sarah rummaged about in her purse until she found a twisted cigarette she had cadged off a stranger on the bus. She went to the fireplace and lit it with a large Blue-Dot match she struck against the brick hearth. She had been looking forward to this cigarette all afternoon.

After a while Raymond said, 'I'm going to tell Grandma you started smoking again.'

Sarah took another long, cloudy drag from the cigarette. She flicked the porous gray ash into her left hand.

'You tell Grandma about my one measly cigarette, sport, and I'm going to chop off both your arms.'

'It is not surprising, I think, that it was the serene and irreproachable Yama, king of the dead and unliving, who first manifested supreme knowledge of the universe and its mysterious Ways,' Mrs Stanford said, beating eggs and milk into a wooden bowl. 'In fact, it is the divine Yama who tells of the great fire that leads to heaven. Salvation is a process of self-destruction. Men die in order to live again. They forget what they know in order to learn what they can never forget.'

Mrs Stanford paused to gaze out the window, which was streaked with mineral deposits and bird droppings. On the stove, butter was sizzling in a saucepan. 'I think there is a peace which we all learn as we get older, Sarah. It is a feeling of peace which young people have a lot of trouble fully appreciating these days, especially since they are so pointlessly preoccupied with sex, drugs and loud record albums. Once you move back with Charlie, you should consider taking Yoga classes. You should learn to meditate, and read Eastern philosophy. You don't have to *believe* in Eastern philosophy like the way you do in, say, most of our money-conscious Western religions. But that doesn't mean you can't learn to breathe properly the way your body wants you to breathe. Instead, you're always hurrying and out of breath, which is just the way capitalist imperialism wants you to be.' Mrs Stanford chopped cheddar cheese on the breadboard, then poured the eggs and cheese together into the sizzling saucepan.

Sarah snapped her eyeliner down emphatically on the table. 'First off, Mom, I'm *not* moving back with Charlie.' She reached for her blush, leaned into the table again and peered at her reflection in a warped shaving mirror, arching her eyebrows one at a time. 'In addition, I'm *not* taking Yoga, and I'm *not* learning to levitate or sing Buddha songs under the Buddha tree. Instead, I am going to find a job that I enjoy, and which does not leave me feeling underused every minute of every day, like being a housewife for a crazy man does. I'm going to buy my own house with a big yard and garden, instead of living in some tiny under-furnished apartment all my life. I'm going to find a man who appreciates me for the strong, loving person that I am, and who doesn't just take me for granted every single day of the week.'

'Know thou the *atman* as riding in a chariot,' Mrs Stanford said, pouring coffee into mugs. The steam uncurled from between her hands while she poured. 'Know thou the intellect as the chariot-driver, and the mind as the reins.'

'Only pour me half a cup, Grandma,' Raymond said. Raymond was still rereading the same *X-Men* comic book. 'Too much coffee makes me hyper.'

Sometimes Sarah felt as if she were trapped in one of those time-space vortices she often read about in paperback exposés of the Bermuda Triangle or Bigfoot. Every morning she took her copy of the Want Ads from corporate office building to corporate office building and filled out lengthy, puzzling application forms which requested information about her minimal education and work history. Then she returned home every afternoon to find Mother and Raymond sitting in their respective sofa and loveseat, reading their respective *Bhagavad Gita* and *X-Men* while the Channel 7 *Eyewitness News* played on the television-stereo console. Sarah would pull off her shoes and collapse on the floor, leaning into a pair of large black pillows still scribbled with the gray hairs of Mother's long-deceased Persian tomcat, Rex.

'I think Sarah's really bushed,' Raymond said. He held the *X-Men* comic scrolled up in his hand. Sometimes he just stared at the same fragment of comic for hours at a time. 'I believe in reading carefully, and rereading carefully as well,' he said, whenever Sarah gave him a critical look. 'If you want to know my opinion, the problem with this culture of ours is nobody *reads* anymore. They take everything for granted. They let the world read *them*. Instead, I think people should force themselves to read the world. I say, stare at something until you understand it. And then stare at it some more.'

Sarah looked up at her son while she rubbed her swollen ankles. Sometimes Raymond didn't seem any more familiar to her than a Martian from outer space. 'Raymond,' she said, 'you're dyslexic. You've been dyslexic ever since you were born. You can't read. You can only look at the pictures.'

Raymond unscrolled his comic and shook it resolutely, like some over-severe grown-up with a newspaper. Finding the segment he wanted, he scrolled it up again. 'Sarah, I think we've had this discussion before.' He made a point of not looking at her. 'There's a big difference between someone who *can't* read, and someone who simply doesn't *want* to.'

By the time Raymond helped her upstairs to bed Sarah felt so tired and depressed she couldn't even go to the bathroom to brush her teeth. She lay down upon the soft bed and gazed at the pale ceiling, which kept going in and out of focus. An incipient migraine banged politely inside her head. When Raymond tried to return downstairs to his comic book, Sarah grasped his thin belt and pulled him back on the bed. 'Sit with me for a little while, sport. Talk to me about something.'

'What do you want me to talk about?'

'I don't know. Anything. Talk to me about anything.'

Raymond looked rather weak and perilous, seated on the edge of Sarah's bed. He will never be an athlete, Sarah thought without remorse. Just like his father.

'Grandma hasn't said a single word all day.' Raymond had begun picking his front teeth with the little finger of his right hand. He resembled some contemplative marsupial.

'She gets like that sometimes,' Sarah told him. She felt herself gliding across the white, stippled ceiling where she discovered vast unexplored territories filled with plains and deserts, beaches and cities. Billions of tiny people moved in them.

'Sometimes, though, she starts humming.' Raymond's skinny body grew abruptly agitated. 'Like this,' he said, clenching his jaws and swaying back and forth, as if trying to pump noise up from his lower intestines. He started to hum. He closed his eyes. Then he stopped humming.

He opened his eyes and looked at Sarah. 'Grandma's trying to contact the Cosmic Ether. That's what Charlie calls it. Grandma's trying to meld herself with the Cosmic Oneness.'

'Don't mention Charlie to me right now, sport,' Sarah said, feeling the surge of curious white bodies above her. Now they were leaning towards her, now they were trying to listen. She heard their white voices, she imagined their most intimate white thoughts. Then, just as abruptly, she fell asleep.

'They're looking for high school and college graduates. They're looking for secretarial and accounting skills. They're looking for young girls with great legs and tight you-know-whats. I'd have a tight you-know-what too if I hadn't wasted the last ten years of my life living with Charlie.' Sarah was opening and shutting cupboard doors, looking for Post Toasties. Raymond sat at the kitchen table with a carton of milk and an empty bowl. His hands were folded on the table. He watched them.

'I wonder if Grandma's meditating is really such a bad idea,' Raymond said. 'I mean, maybe truth isn't something solid. Like this table, say.' Raymond's thin, freckled hands grasped the table edge. 'Or this bowl.' Raymond lifted the bowl on his fingertips as if it were an object of sacrifice.

'Here, this is all I found.' Sarah put down an unopened box of Quaker Oats on the table, sat down and lit a cigarette. She felt like tearing the cigarette into tiny little pieces and flinging them into Raymond's face.

Raymond picked up and examined the box. The smiling Quaker Man on the box seemed to fill Raymond with a vague anxiety. 'You're definitely pre-menstrual, Sarah. I don't know if this is the best day for you to go looking for work. I know *I* wouldn't hire you. And, if you'll remember, *I'm* your son.'

'It's not like because you're thirty-two means you're over the hill or anything. I mean, we can't all be sixteen. Do you see those little girls they've got on the covers of the women's fashion magazines these days? They're all fourteen years old. I think that's sick. If you ask me, that's exploitative and sick.'

Raymond showed her the box of Quaker Oats. 'You're supposed to boil this stuff for like fifteen minutes or something. And then it tastes just like glue you wouldn't feed your dog.'

When Sarah went job hunting later that morning even the weather seemed menacing and austere. The sky was white, flat and dimensionless. There weren't any birds in it, or any planes. Sarah took a sequence of buses into San Francisco, the morning *Chronicle* resting on her lap like a small pet. She couldn't concentrate on the Classifieds; she couldn't concentrate on the news; in fact, she could barely concentrate on her horoscope in the Living Section. Something about Jupiter in her house, opportunities making themselves known, not encouraging familial confidences. Sarah fidgeted in her window seat and noticed a hard fragment of something on her left stocking. She scratched at it, and the stocking began to run. Everybody on the bus seemed to be taking turns watching her, as if they were keeping a sort of communal log. Sarah got off the bus at Van Ness and stood slightly dazed and obvious on the sidewalk. Then she remembered she had left her newspaper on the bus. The bus was taking off, and an elderly woman at the window was gazing vacantly at her, as if trying to remember something that had happened to her over thirty million years ago.

'Many of our employees begin as junior clerks,' the young man in Personnel told her. He held Sarah's application form in front of him while he talked, as if he were reading from a script. 'In fact, there are many executive officers in our company who began in Customer Service, from where they were eventually promoted to high-level managerial positions in marketing, sales promotion, advertising, and even group insurance. Junior clerk is a part-time position, but you could be elevated to full-time employment within the next, say, twelve to eighteen months. This will make you eligible for our group medical and dental plan, which is, of course, one of the most extensive health plans available in the entire Bay Area. And then, of course, we provide group discounts for all of our employees to local film theaters and shopping centers. We are, of course,' and at this point the young man cleared his throat and gave the piece of paper in his hand a succinct little shake, 'non-union. But we like to think we take better care of our employees than any trade union ever would.'

*

When Sarah returned home that afternoon she found Raymond listening to scratchy Sarah Vaughan albums on the stereo-console. Raymond was sitting on the floor dipping frozen strawberries into a glob of sour cream and brown sugar on a green ceramic plate. Sarah could still remember the day her mother received that gratuitous plate in exchange for purchasing a full tank of gas at the Esso station.

Raymond was watching television with the sound off. An attractive young man was seated among the bright paraphernalia of a cardboard spaceship cabin. He was wearing a bright, aluminum-like spacesuit. A small fluffy green ball with eyes, like the charm on a keychain, was perched on his shoulder. Raymond sucked on a frozen strawberry, his mouth rimmed with red pulpy juice.

'Did you get a job?' he asked.

Sarah placed her purse on the coffee table, she sat down and pulled off her high heels. She rummaged in her purse for the pack of celebratory cigarettes she had purchased from a newsstand.

'I think I did,' she said. 'I think I really did get a job. Where's Grandma?'

'I don't know.'

'What do you mean you don't know?'

'I think Grandma went away on a retreat.'

'What do you mean Grandma went away on a retreat?'

'Why do you keep asking me that?'

'Why do I keep asking you what?'

'Why do you keep asking me what I mean? Like I said, I *think* Grandma went away on a retreat. There was a large white van that came to our door during *Dialing for Dollars*. These two young men got out. They were both wearing these bright white T-shirts that said Karma Kounselors on them. They knocked on the door and asked if Grandma was home. I went upstairs and got her. She was sitting on the floor of the master bedroom chanting her mantra, over and over again, I guess so she wouldn't forget it. I told her it was two young guys in a white van from Karma Kounselors, and she opened her eyes. She went to the closet for her coat, and gave me all the cash in her purse. Then I followed her downstairs and watched her get in the van. Before they closed the back doors of the van, the two young men talked to her for a few moments. They pointed at me a couple of times, where I was standing at the front door. Grandma shook her head. Finally they shut the back doors of the van. When the two young men got back into the front seats, the young man closest to me waved and smiled. They seemed like very nice guys. When Grandma gave me the cash from her purse, she told me she was going on a retreat. She was withdrawing from worldly congress. She needed to get her head together, and gave us free rein of the house. No matter how many times she asks, we're not supposed to send her any more money, or any of the credit cards she left in the top drawer of her bureau. Now, there you have it. You know everything I know.'

Raymond showed Sarah his plate. 'Do you want some frozen strawberries? I went to the grocery store after Grandma left. We were down to our last slice of bread.'

That night, Sarah lay down on Raymond's couch while Raymond watched rock videos on television. The house seemed thin and unresonant – the marred furniture, unravelling draperies and antimacassars, chipped and nicotine-stained porcelain figurines.

'I've lived here ever since I can remember,' Sarah told Raymond. She sipped intermittently from her rum and coke. 'Just your grandmother and I, except when your Great-uncle Teddy came to stay with us. Your Great-uncle Teddy never worked a day in his life, and he came to live with us after his first wife, Rose, threw him out of her house. Grandma always referred to Uncle Teddy as 'the man who came to dinner,' which is a reference to a famous play about a man who comes to dinner at a friend's house and then doesn't leave. That is pretty much what your Great-uncle Teddy did, which is to say he came to dinner one night and then he didn't leave. He slept right here on your sofa, and never contributed one nickel to pay for groceries or house maintenance, and never washed a single dish.'

On the television, an all-girl rock group was singing a song about partying at the beach. While they sang, large cardboard cut-outs of girls waved their arms and danced on a cartoon approximation of a beach. Raymond was lying on his stomach, his arms folded around his head. He snored slightly, like some small hibernating mammal. A squirrel, perhaps, Sarah thought. Or some thousand-year-old desert tortoise.

'Eventually your Great-uncle Teddy married another woman who was almost as big a sucker as his first wife. He went to live with her in her house in Gilroy. She was a registered nurse, with a small income from the estate of her parents. Your Great-uncle Teddy moved in with her like a shot. He took all his clothes, and some of our furniture. Every year or so he would do us a big favor and come over for a visit. Then he would sit on this sofa right here and drink our booze and eat our food. If I get this job, I am going to buy us a house where your Great-uncle Teddy will be strictly off-limits. In fact, men in general are going to be pretty much off-limits in our house, unless, of course, they have enough money to pay their way.

'We got a letter from Charlie today,' Raymond said the next morning, serving Sarah something on a plate. Sarah had overslept and now it was nearly ten a.m. Raymond was holding the smoking pan in one hand; his other hand wore a charred oven mitt. He nodded in the direction of Sarah's plate while Sarah lit her first cigarette.

'I figured you should have a hot meal or something,' Raymond said.

Sarah took a drag from her cigarette and placed it delicately on the rim of a saucer. Tentatively, she tapped the top slice of burnt toast on her plate with a butter knife.

'What's this?' Sarah asked. 'It looks like chili between two pieces of burnt toast.'

'It's a chili and cheese sandwich,' Raymond said. 'I kind of invented it.'

'What did Charlie say in his letter?'

'I've got it here if you want to read it.'

'I don't want to read it. Just give me the gist of it.'

'The gist of it is that Charlie says he really misses us and wants us to come home.'

'He wants us to come home so he can finish the job of stabbing me to death with a kitchen fork.'

'He was just gesturing, Sarah.'

'And I guess he was just gesturing the time before when he tried to burn my face off with a pan of boiling tomato soup?'

'Sarah, you know as well as I do that Charlie is not a mean or vindictive man. He's just clumsy. How do you like it?'

'How do I like what?'

'Honestly, Sarah. You don't listen to a single word people say to you, do you? How do you like your chili and cheese sandwich?'

'Oh.' Sarah tapped the burnt toast twice with her butter knife. 'It looks perfectly fine, I guess.'

*

Three days later Sarah started work at the *San Francisco Tribune*, where she attended a week-long training seminar in a long glass-walled board room and learned to institute changes in customer service. Sarah's Training Service Coordinator was a young, freckled and widely bespectacled man named Bill Babbitt who always wore a nice suit and tie, and who enjoyed comparing almost every aspect of Customer Service to his favorite sport, which was skiing.

'Much like riding a mogul,' Bill might say, his teeth sparkling in the early morning fluorescents while Sarah and the other trainees frowned into their Styrofoam coffee cups, 'you must learn to anticipate the customer's every swerve and jump. You must be prepared to suggest possible ways in which even the customer's unvoiced expectations can best be met. Now, Sarah, my name is Mrs I. M. Newsworthy, and I have just called you at your Service Response Position. I want to initiate a temporary stop on my newspaper. I am leaving for vacation on Friday, and returning three and one half weeks later. Which form do you reach for?'

'The red one,' Sarah said.

'The Permanent Stop form? But didn't I say I just wanted a *temporary* delay in service?'

'Yes, but this way the Agent receives a stronger reminder to resume customer service than he would, say, if we filed the yellow Temporary Change-of-Service form.'

Bill Babbitt smiled; his white teeth flashed around the room, one trainee at a time, as if dispensing a ration. 'That's absolutely right, Sarah. That's absolutely, positively correct.'

'Usually Bill calls on me to answer the harder questions,' Sarah told Raymond during the second week of training. 'This is because, unlike many of the younger, less experienced girls, I am always paying close attention in class. Your grandmother was very disappointed when I had to leave high school due to pregnancy. This was of course primarily due to the fact that Charlie couldn't support us on his miserable salary.'

Raymond had grown rather gray and desultory over the past few days. He was lying on the couch with his latest acquisition from the drugstore newsstand, an automobile customizing magazine, and twirling a curl of auburn hair behind his pale ear. He resembled some forlorn Romantic poet in a museum painting. Sarah reached over and checked his high forehead for fever.

'Sarah, how come you never married Charlie?'

Raymond's forehead felt cool and dry. The absence of symptoms made Sarah uneasy.

'Why don't you stay home tomorrow, sport? Sit around and relax and watch television.'

'Sarah, I sit around every day and watch television. It's like all I ever do.'

'Well, sit around and watch TV tomorrow, too.'

'School's out, I don't know anybody in this entire neighborhood, and there's nowhere I can walk except to the supermarket to buy these lousy magazines I can't even read.' Raymond flung his tattered issue of *Customized Car Magazine* at the floor. 'I probably will sit around tomorrow watching television. I bet that's exactly what I'll do.'

'I thought you always said you just didn't *want* to read.'

Raymond sighed. 'Whatever.'

Sarah reached out and touched Raymond's forehead again.

'You're definitely starting to feel a little warm,' Sarah told him.

'Who cares,' Raymond said.

On Friday Bill Babbitt showed them an inspirational corporate film about fifty years of public service at the *San Francisco Tribune* which featured guest appearances by many easily recognizable television and film celebrities, including Ed Asner, the actor who played newspaper editor Lou Grant on the recently cancelled TV series. When Sarah returned home that afternoon she found Raymond flushed and sweaty, covered over with the heavy duvet from his grandmother's room, and peering at a punk-rock video on MTV with the sound off.

'How do you feel, honey?'

Raymond held the edge of the duvet up to his nose, as if he were defending himself from toxic fumes. 'Grandma called,' he said, his voice muffled by the blanket. 'She says she forgot her Check Guarantee Card.'

'Did she say when she was coming home?'

Raymond coughed. 'She said when her Deep-Breathing Counselor came through town tomorrow he'd stop by and I should give him her Check Guarantee Card. She wanted to know when we're moving back to Charlie's house, but she said we're welcome to stay here as long as we want.'

Sarah slumped into the stern Barcalounger. A pair of identical, beautiful young blond women on TV were comparing different brands of roll-on deodorant by writing their names with them on a sheet of plexiglass. 'Let me think about it,' Sarah said after a moment. Then she leaned over and checked Raymond's forehead again.

'You're burning up, honey,' Sarah said. She held her hand against Raymond's forehead for quite a long time. She thought about a slick pictorial advertisement in a women's magazine she had seen that day during lunch break. In Bermuda the water was translucent and blue, and women displayed their long tan legs against white sandy beaches. Raymond didn't stir. His eyes closed slowly.

'If the guy comes by tomorrow for Grandma's Check Guarantee Card,' Raymond said dimly, drifting away into his own blossoming temperature, 'I'm not giving him any Check Guarantee Card. I'm not even going to answer the door. To be perfectly frank, Sarah, I think Grandma's got herself messed up with the wrong crowd entirely.'

The last time Sarah Stanford had entered a hospital she was visiting her father at the Veterans' Ward in San Francisco County where he lay dying of prostate cancer. Sarah's father had shared a long, wide intensive-care unit with a variety of other semi-conscious elderly men, a flock of hurrying white nurses with stiff angular white hats, and a flat, medicinal odor that resembled cheap deodorant or children's toothpaste. Sarah's mother had held Sarah's hand and escorted her down crowded aisles of hard white beds and tall plastic partitions. The gray irresolute faces of old men gazed at Sarah and her mother as they passed with a strange, urgent expectation. Then her mother brought her up to one particular white bed with an unconscious old man lying in it. The old man didn't move or open his eyes. Sarah couldn't even detect his breathing.

'This is your father,' Sarah's mother told her. 'He is very ill and hasn't much longer to live. But I thought you should at least have a chance to see him before he dies, and give him this opportunity to make amends. He was a totally useless rat while he was alive, but now that he's dying he deserves to be forgiven.'

*

When they checked Raymond into St Luke's Hospital Sarah was asked to wait in the lobby while they ran tests on the third floor. She was so nervous and emotionally distraught that she bought a pack of Marlboros from the Gift Shop and chain-smoked hastily. She ate a packaged cheese sandwich from the Visitors Cafeteria, two bags of M&M chocolate-covered peanuts, and a box of Milk Duds that was originally intended for Raymond.

'Where's my Milk Duds?' Raymond asked, prone in his wide hospital bed. Tubes were taped to his flared red nostrils; his brow was sweaty and rough with shiny pimples. He seemed very small and defenceless in his wide bed.

'I ate them,' Sarah said, rummaging in her purse for the cigarettes.

'*I'm* dying and you ate *my* Milk Duds?'

'You're not dying, sport, and I didn't exactly eat *your* Milk Duds. I'll get you some of your own Milk Duds before I leave.'

'I've never been in a hospital before,' Raymond said, gazing around him at the other children who lay insensibly attached to various incomprehensible tubes and machinery. The other children were surrounded by large elaborate stuffed toys, picture books and board games. Battleships, Monopoly, The Game of Life. 'They treat you really nice here,' Raymond said. 'And some of the nurses are really pretty.'

'You have to stay here overnight so they can run more tests.' Sarah struck three matches before she managed to ignite her trembling Marlboro. 'Oh, and here. I brought you something.'

Raymond took and unscrolled it with intense concentration, as if he were reading some dire Roman prognosticae concerning slavery or war.

It was Raymond's *X-Men* comic book.

The following morning Sarah had to wake at five-thirty and catch the six-fifteen bus at El Camino Real. She felt bloated and immense with all the beer and cigarettes she had consumed the night before. With increasing reluctance, she examined herself in the mirror while she brushed her teeth. Her face was puffy and slightly flushed, her eyes dry and bloodshot. Every time she thought about Raymond in his wide, lonely hospital bed she thought she was going to start crying, but then, with a strange feeling of inadequacy, all she kept thinking was, if I'm not in the shower in the next two minutes, I'll be late for my first day at work.

She was seventeen minutes late, and before she could even locate her personally labeled headset in the Employee Equipment Room, she heard her name being called over the departmental loudspeakers.

'Will Sarah Stanford please report to her Group Training Coordinator in Cubicle number Six? That's Sarah Stanford, Cubicle number Six, please.'

Cubicle number Six belonged to the bank of glass-walled offices in the center of the Department of Customer Services. Across the broad, thinly carpeted floors of the Department, Customer Service Operators sat in groups of six at their Pod Performance Modules – tiny desks arranged around an asterix of sound-insulated pasteboard partitions. The operators all wore identical headsets and spoke into tiny V-shaped glass mouthpieces, like cyborg-fitted astronauts on a spaceship.

'Much like running a downhill performance course,' Bill Babbitt explained, 'speed and skill are pretty useless if you suffer a late jump from the starter's box. Now, I realize that these are pretty difficult hours to get used to, but in the future, whenever you are late and I am monitoring attendance, I'll have to fill out a green Employee Service Report, whether I like it or not.'

Bill Babbitt picked up one of the green forms from his desk and waved it at her, emphasizing its severe reality.

'Once three Employee Service Reports have been filed, the Clerk receives a formal written admonition. After such an admonition has been made, the Clerk is placed on parole, and can be dismissed at any time according to the Service Coordinator's personal discretion. Now, I don't think either of us wants to see you dismissed, do we, Sarah? Especially since you seemed so promising during last week's Training Placement Seminar.'

For the rest of the day, Sarah felt she was running systematically late for everything. She had trouble attaching her headphones to the switchboard, she kept running out of red Stop Service Reports, and then, even though she was dying for a cup of coffee, she accidentally worked through her entire fifteen minute break. What was worse, almost every customer who called was rude, abusive and impatient.

'Good morning,' Sarah said, as cheerfully as possible. She was usually trying to disentangle herself from the coiled telephone wires, or rub the ink stains off her fingers with a paper tissue. '*Tribune* Customer Service, Sarah speaking. How can I help you?'

'Well, you can start by getting me my goddamn newspaper. It's nearly ten in the morning and I still haven't received my goddamn newspaper yet.'

'Would you like to give me your address?' Sarah asked, reaching for a white Service Report form. 'And I'll relay your problem to the local delivery agent as soon as possible.'

'I already gave you guys my goddamn address over an hour ago, and I still haven't received my goddamn newspaper yet.'

When Sarah got off at one, the Number 42 was pulling away just as she arrived at the bus stop, and even when she rapped on the accordion-like door with her fist the driver refused to stop. So then Sarah had to sit in the hot sun for almost thirty minutes before the next bus arrived.

'You'd think for a woman who'd worked all morning, the bastard would have bothered to stop.' Sarah was still rehearsing her anger an hour later, pushing Raymond down the hospital corridors in a wheelchair. 'I mean, here I am, a single woman with a child in the hospital, and he couldn't wait one lousy little second. What does he think – that I've got all day to sit around waiting for the next bus to come? Doesn't he think I might be worried sick about my only child, who's having blood tests, for chrissakes?'

Sarah released her anger in short, hard little bursts. Just enough to relieve the pressure, but not enough to lose the edge.

Raymond was hunkered down in the wheelchair, his chin pressed against his thin, sunken chest. He glanced around ruefully at the nurses and orderlies in the hall, as if anxious about paparrazzi. He was flushed, and wearing a disheveled white hospital robe which didn't cover his bare, skinny white legs. 'I don't see why I have to be pushed in a wheelchair, Sarah. I look like some sort of geek.'

'Hospital policy,' the nurse said, walking along beside them. 'Just relax and enjoy the ride.'

In the taxi on their way home, Sarah said, 'So then he gives me a lecture about being late and being responsible and not letting the team down, just like I was a little girl in school, and then I missed my coffee break, and the people that *call* you at that place! They are totally, totally horrible absolutely *all* of the time. In fact, they were so horrible I even took down some of their addresses. What should I send them, sport? I think I should send each and every one of them a nice little surprise in the mail.'

Raymond was wearing his blue padded nylon jacket over his short white hospital robe. 'You're going to be pissed off, Sarah, but I called Charlie today. He even came by to visit when you were at work.'

'You what?' Sarah felt dazed and breathless for a moment. She looked at Raymond. 'You did what? What did you do?'

Raymond shrugged. He gazed out the window at the Veterans' Cemetery – rows and rows of white, iconic tombstones and soft green grass.

'I know you don't believe me, Sarah, and I'm not trying to scare you or anything, but I'm dying, and I needed to talk to someone. I don't need to wait for the test results to know it, either, because I've already known for weeks. I've been getting this old lady at the library to help me read up on my symptoms in the Home Medical Encyclopedia. I've got all the symptoms, Sarah – they're obvious as a heart attack. First, there's these enlarged lymph nodes under my arms and around my neck. My liver and spleen project from under my ribs like the edges of plates. I'm tired all the time, and I've always got the sniffles, and I've generally got this really bad attitude towards life, which is probably the only part about leukemia that seems perfectly normal to me. I've been studying about leukemia so much in the past few weeks, Sarah, that I have literally scared the living piss out of myself. I keep intending to tell you about it, Sarah, but frankly, you're a lousy listener. And then even if you *did* listen to me for one *single* second you'd probably just tell me I was nuts like you always do. But I'm not nuts, Sarah. I wish I was, but I'm not.'

Sarah couldn't stop staring at Raymond. She just couldn't figure. This was her child. It had come from her body. It was equipped with its own internal organs, its own brain, its own cells and arteries and oxygen and blood. Biology was difficult enough to comprehend, but here, as if that weren't enough, Sarah had to confront this wide interior subjective life of Raymond, staring out at her from behind Raymond's face. Where the hell did he come from? Sarah thought. What parts of my body made him? What the hell was I thinking about when it happened?

Sarah said, 'But you *are* nuts, sport. You're completely whacking bananas.'

Raymond looked up at her again. His face was filled with color. He even looked like he had put on a little weight while he was in the hospital.

'I knew you'd say that,' Raymond said.

'You're a hypercondiac,' Sarah said in the same clipped, mechanically estranged voice, as if she were reciting her grocery list into a dictaphone. 'I've got to start making more money so I can get you visiting a psychiatrist. And I mean right away.'

'That's hypo*chon*driac, Sarah.'

'Same difference.'

'And I'm not a hypochondriac, Sarah. *You* are. You're the biggest hypochondriac I ever met in my life.'

With a faint, polite lurch, their taxi arrived at the curb.

'That's seven fifty, please,' the driver said.

When Sarah paid the driver and helped Raymond climb out of the cab, she looked up and saw Charlie's car parked in the driveway. A '73 Chevrolet Impala with peeling vinyl roof, bad paint and a severely crumpled front end. The doors and hood were marked with severe pits and dings, as if someone had been taking target practice at it with a B-B gun. The dashboard was sun-warped and discolored, and the torn plastic seat covers exuded white cotton batting and yellow foam insulation. The back seats were filled with various spare auto parts – rusty wheel rims and hubcaps and oily carburetors and black, sooty mufflers.

Charlie was sitting on the front porch, wearing blue overalls. He had shaved and combed his hair, but he was still wearing those horrible blue overalls.

'Hi, Sarah. Hi, Raymond.' Charlie was sitting with his hands pressed between his knees.

I always hated that damned car, Sarah thought.

'I don't want to listen to you anymore, Charlie,' Sarah shouted through the bolted door of her bedroom. 'I don't want to hear your excuses, I don't want to clean your smelly underwear, and I especially don't want to see your stupid face ever again. I've got my own life to live now, Charlie, and there isn't anything you can do about it. I've got a job where everybody likes me, and a lot of new friends, and a whole range of career opportunities to choose from. And pretty soon, once those payroll checks start rolling in, boy, I'm going to get my own apartment, and my own car, and do whatever the hell I want with my life. I'll buy my own house, or even start my own business. I'll meet some halfway decent men for a change, or maybe just go into celibacy and save my creative juices for more important self-affirming activities. You never really appreciated me, Charlie, because I had ambitions about life and you didn't have any. All you ever wanted to do was work, drink beer, and sit on the sofa and watch that stupid TV, and I've had all I can take, Charlie. All I can take by a long shot.'

'Why don't you let me fix you some dinner?' Charlie said. Sarah could feel his shoulder impressing her thin plyboard door-frame. I am here, the weight said. I am here and you are there.

'Don't try to make me feel guilty, Charlie, because that won't work this time. I *knew* you'd do that. I *knew* first chance you got you'd try to make me feel guilty again.'

'How about soup and sandwiches, honey? Peanut butter and tomato and lettuce. Just the way you like it.'

'I'm not listening, Charlie. I've got my hands over my ears and I'm not listening.'

'Or even a pizza. I could order a pizza with ham and pineapple from Number Uno. With extra mozzarella.'

'Stuff you and your pizza, Charlie. Now go visit your son who adores you so much and leave me alone before I call the cops. Go visit your son you obviously love so much since you let him be born out of wedlock.'

'But, *Sarah.*' Sarah heard the wide slow fall of energy in Charlie's voice. '*You're* the one who didn't want us to get married.'

'*Ha*, Charlie. *That's* the best joke I've heard all week. You know as well as I do that if you'd *really* wanted to marry me, you wouldn't have taken no for an answer.'

Eventually, standing alone in her dark bedroom, Sarah felt Charlie's pressure withdraw from the door. She heard the low solid slump of him retreating down the hall, the thaw and muddle of Charlie's mind turning. Hearing Charlie's mind at work was like hearing the freezer defrost. It emitted thick clumps and bright, sudden cracks in the night when you least expected them.

She could hear them now conferring in the bright kitchen. Their voices were low and assured. They were men, and they knew what each other was all about. They liked ball games on the radio, they liked beer, they liked looking at girls even when those girls had no intention of looking at them. Men simply didn't make sense, and they worried Sarah constantly. Even when she wasn't thinking about men, somewhere in the back of her mind she was always worrying about them.

'Isn't it always,' Charlie's voice said, bobbing to the surface of their conversation like a porpoise. 'Doesn't mean anything. How about more soup.' And then Raymond saying: 'Sarah loves you, Charlie. I know you can't always see it, but in her own way I think Sarah loves you very much.'

'It's not leukemia,' Sarah said, staring at the same cigarette, preparing to strike the same paper match. 'I don't care what they say. Doctors don't know everything, Raymond. Trust your mother on this one. Doctors just want to make you *think* they know everything. That way you'll pay your bills. That way you'll help them pretend they're God's greatest gift to medicine.'

'It's called acute lymphoblastic leukemia. Don't go fleeing into denial on me, Sarah. I'm scared enough as it is.'

'I'm not de*ny*ing anything, Raymond.' For the first time since she hung up the phone on Raymond's doctor, Sarah looked at her son. He was busily applying butter and syrup to his Toast-Up waffles. He looked better than he had in months.

'It's because I'm pumped full of somebody else's blood,' Raymond said, reading her mind as he reached for the Tang. 'My luck, I'll probably catch AIDS or something.'

'Don't say that word in my house,' Sarah said. A keen adrenalin rush reached through her body. She sat up straight.

'What word is that?' Raymond asked. He was looking over Sarah's shoulder at the floral-patterned wallpaper.

'You know what word I mean, sport. *AIDS*.' Sarah was suddenly struck by the notion that somewhere deep inside herself she had just come to an important decision about something. It was only a matter of time, she thought, before she recognized what that decision was.

'And don't accuse me of denying anything, Raymond. I'm just telling you what *I* think, that's all. It's not like you're the only person around here whose opinion matters.'

All day long Sarah felt cellularly lapsed and momentless, as if time and space were racing formlessly around her.

'Frankly, Sarah, I don't think this is exactly the perfect day to discuss increasing your duties here at the Customer Service Center.' Bill Babbitt was sitting in the Central Coordinating Module with his feet up on the monochrome black desk. On the monitoring console, a bright digital clock ticked ruthlessly. 'If I can be blunt with you, Sarah – and I know you're the sort of person that prefers to be dealt with frankly – we haven't even decided if we're keeping you on the morning shift or not. You were tardy without prior notice on Tuesday, and even with prior notice you were nearly three hours late today. Now, I understand all you've tried to explain about your new living situation. Your split with your boyfriend, your little boy's disease. And I sympathize with you, Sarah; I really do. But you can't run a business by sympathizing with people, can you? I think you'll agree there's a lot more to running a successful business than liking people, and hoping they like you.'

'Now look,' Sarah said, leaning earnestly forward in her creaky plastic chair. 'I don't like to ask favors, but it's not like the extra hours I need – it's really the medical insurance. I mean, if it turns out Raymond needs a lot of, you know, treatments or something. I just realized this morning I'm not covered for a single thing. And neither is Raymond's father, who sells used and rebuilt auto parts out of the back of his car.'

For the rest of the day Sarah rehearsed her frantic disapproval of men. Under her breath she muttered things like, 'Men are, without a doubt, the most miserable creatures who ever set foot on the face of this planet.' Or: 'Look, sport, I don't care if you grow up to be a pornographer or a junkie. Just don't grow up to be a man, okay?' On her way home, she sat on the bus bench crying for nearly half an hour, drenching every crumpled Kleenex and fragment of Kleenex she could dredge from the depths of her large ersatz-leather handbag. Cars and motorcycles roared past, overhead power lines spluttered and hissed. Whenever Sarah reached an emotional lull, she told herself out loud, 'Screw Bob Babbitt. Screw Charlie. Screw my fucking father. Screw the fucking president of the fucking United States.' Then, almost as an afterthought, she succinctly began crying again.

'Screw Bob Babbitt,' she said again, like a litany or a prayer. 'Screw Charlie. Screw my stupid fucking father. Screw the fucking goddamn president of the fucking United fucking goddamn United fucking States.'

When she looked up again, she saw a hairy, ectomorphic man in filthy matted clothing carrying a sign towards her. The sign said:

LESS LUST FROM LESS PROTEIN

Less fish, eggs, beef, pork,
nuts, dairy products, coffee, sugar, tea

EQUALS
less labor, power, capital, fuel, war, poverty,
disease, love, suicide, heartache, muscle, bangs and pops

LESS LUST MEANS MORE PRODUCTIVITY
MORE PRODUCTIVITY MEANS MORE PAY
MORE PAY MEANS MORE ENERGY FOR OUR PERILOUS
SPACESHIP EARTH

'I don't think I understand,' Sarah told the man. 'Are you saying we should eat more protein, or less?'

'That depends on whether you want to increase your lust-potential or not,' the man said reasonably, and scratched his belly with long callused fingernails. He smelled like sour cheese.

Sarah thought about this for a moment.

'I guess I don't really care that much one way or another,' she said.

When Sarah returned home that afternoon she found Raymond sitting on the living-room sofa. Wearing only his beaded blue flannel pajama bottoms and white wool tube socks, Raymond looked severely pale and underinflated, his lips moist and swollen, as if he had expended his transfused energy sucking hard candies and lollipops. His body gave off a thin, languorous heat, like the mist generated by a humidifier.

'I didn't want to let him in, Sarah,' Raymond said, indicating the large man seated beside him on the sofa. 'He kind of insisted, like.'

The man was extremely tall, fair-skinned and thirtyish, with long blond mossy dreadlocks and bifocal granny-glasses. His sunlamp-tan was redolent, reflective and hard, like the exoskeleton of an insect. He had enormous hands with blunt, phallic fingers, and a wide smile filled with bright, porcelain-capped teeth. He was wearing blue straw-base sandals, loose-fitting beige linen trousers, and a neatly ironed white T-shirt which said: KARMA KOUNSELORS – DR LEWIS – HEAD HONCHO.

'I'm afraid your son suffers from a vertical imbalance in his positive and negative ion levels,' Dr Lewis said, with a smooth blue resonance of authority. He spoke evenly and without embarrassment, as if he were reciting weather indices to an empty television studio. 'He is centered neither spiritually nor bodily, and as a result he doesn't feel involved in the vaster totality of nature. I suggest that first we teach him to breathe properly. *Pranayama, Ujjayi, Khumbhaka.* A body capable of renewing itself with *pranic* energy becomes a source of healing not only to itself, but to others as well.'

'He was looking for Grandma's Check Guarantee Card,' Raymond said, watching Dr Lewis's hand pet his brow as if it were a particularly loud wasp. 'Then he started going through her drawers and closets. He wanted to go in your room, but I told him I'd call the cops. He's got a pair of Grandma's earrings in his right-hand pants' pocket. He thought I didn't see him take them, but I did.'

Dr Lewis smiled, closed his eyes, and rested the backs of his hands on his knees. Perfunctorily, he hummed. Then he said, 'Mantric yoga unifies the body with the voice, the spirit with the flesh, the yang with the yin.'

'Sarah,' Raymond said. Raymond looked impossibly tiny and frail beside Dr Lewis. 'Could you please get this guy out of here? He really depresses me.'

'One minute, sport,' Sarah said.

She took her handbag with her into the kitchen and hid it underneath the sink in a red plastic pail. She opened the utensils drawer and rummaged through a mismatched assembly of knives, forks, can openers and plastic bottle caps. Then she tested the weight and grip of a wooden ice pick. The metal blade was rusty, and the wooden handle scored with black, greasy cracks. After a moment's hesitation, Sarah returned to the living room and showed Dr Lewis the ice pick.

'Get out of my house,' Sarah said. 'And take your load of Buddha bullshit back to summer camp.'

Dr Lewis continued to smile. He contemplated the ice pick in Sarah's hand. His entire presence radiated peace and tranquility and images of sunny river banks filled with green fragrances, honey-colored sunlight and gleaming silver fish.

'Why are you so angry, Sarah?' Dr Lewis said. 'Don't you understand that anger divorces you from worldly harmonies? Why don't you put the ice pick down, Sarah. Come sit with me and your son on the couch.'

'Personally, I don't mind anger that much,' Sarah said. 'Anger keeps me awake.'

She showed Dr Lewis the ice pick again.

'Now, if you don't get out of here in exactly two seconds, I'm going to insert this thing through both of your gonads.'

After Dr Lewis departed, Sarah fixed Raymond and herself Kahluas with coffee. Raymond took two sips and was immediately high.

'I really liked that line, Sarah,' Raymond said. He still seemed perilously thin and deflated, but his face was already filled with color again. He was pursuing the cracked remnants of yellow, parchment-like Pringles at the bottom of a long tubular container, like a kitten pawing the last drops of cream from a bottle.

'What line was that, sport?' Sarah said. She pulled off her shoes and sat beside him on the couch. Now they were both gazing into the deep, expendable light of the television.

'You know. The one about piercing his gonads.'

'He just better not try messing with me again,' Sarah said. She had already finished her drink, and was lighting another cigarette. The ice pick was sitting on the coffee table beside her coffee saucer. 'In fact, nobody better try messing with me ever again. I'm righteously pissed off. I'm so sick of men, I've decided to become a lesbian or something. I think you should too, sport. I think we should become lesbians together.'

Sarah felt Raymond's customary disaffection enter the room and sit down beside her on the sofa like a large beast. He scowled.

'That's totally disgusting, Sarah. You make me sick to my stomach.'

'I must have gotten it from your Grandma,' Sarah said, and poured them both another Kahlua. 'You think your Grandma's something of a nut-case now, then you should have seen her when I was growing up, boy. All she did was booze and bring home strange men. The first time I took Charlie to visit your Grandma? She was drinking Southern Comfort like it was going out of style, and telling Charlie about all the back-lot abortions she had when she was young. I was so embarrassed I went into the bathroom and threw up. Then Charlie, the biggest genius of the twentieth century since Einstein, tells me all the way home what a terrific mother I've got, and how much fun she is, and so on and so forth, and how I should appreciate her, and have her over more often, and on and on and on. Your Grandma and Charlie, what a combination. It's a wonder I didn't go completely nuts and take you with me.'

Raymond was asleep, wrapped around a small warm pillow. Sarah leaned over and kissed him on the ear.

'Do you want to sleep upstairs with me, sport? I don't mind.'

Raymond scowled. 'Don't be ridiculous.'

Sarah covered him with a blanket and took her Kahlua upstairs. She was thinking, $347, $694, $1388 by September third. I'll pay all the bills out of Mother's checking account. Radiotherapy treatments at $250 apiece, and then hospital bills, doctor's fees, blood transfusions, plasma. Sarah flipped off her shoes and lay down on her bed. Medical insurance, full-time benefits, night-work, a second-hand car. Then her own apartment some day, if only she could get ahead. The light was on over her bed, and she felt herself descending into fathoms of negative numbers. Minus five hundred, seven hundred, two thousand, twenty-three hundred. Three thousand, thirty-five hundred, four thousand, five. She lay motionless on her bed. She couldn't move. She heard soft footsteps on the carpeted stairs. The closet door opening, closing again. She was uncertain whether her eyes were open or not, but she could still see the light, white and powdery, like the fragrance of blossoms.

The ice pick, she thought. If she could move a muscle, she could reach for the ice pick on her bureau. Then she heard the door open. Then she heard his low voice speaking.

'I changed my mind,' Raymond said. He turned off the bedroom light, climbed across the bed with brutal awkwardness, and pulled himself underneath Sarah's transferred warmth in the blankets.

'And by the way, Sarah. Don't forget,' Raymond said, just as Sarah began descending into the dark numbers again. 'That guy's still got Grandma's earrings.'

In the morning they received a letter from Grandma. The letter said:

Dear Raymond and Sarah,

How are you? Things here at the summer camp are just wonderful, and I'm on a diet. How about that?

Whatever you do, and whatever Dr Lewis tells you, don't give him any money, or let him bother you in any way, though I would like someday for Dr Lewis to teach Raymond how to breathe properly, since he is such a deeply nervous young boy almost all the time, and has absolutely no idea how to keep his inner polarities balanced.

Meanwhile, as to yourself, Sarah, you are my only child which makes you of special importance to me. Often here during my meditation or counseling sessions, I think back to all the terrible traumas you must have suffered having a miserable screwed up cow like myself for a mother. Never knowing what sort of crazy insane man you'd find eating breakfast in your kitchen every morning, for example, or being embarrassed about what the neighbors said about the crazy goings on at our house, and so on and so forth.

Anyway, I certainly wish I could make up for all the traumas I caused you in your life, Sarah, but frankly I've got so many traumas to take care of in my own life that there's a good chance I may never get around to helping you with yours. Basically I guess all I'm saying is I wish you all the luck in the world in your new life away from Charlie, and feel free to stay in my house while I'm gone, and I can probably help you out with a few dollars every now and again.

I wish I could do more, but I'm afraid I'm pretty wrapped up in my own spiritual growth processes right now, and you're not particularly interested in what I have to say about that stuff, anyway. We all have to muddle through the best we can, I guess, and if I had time to make up for all the crummy things I've done to people in the past then that would be just peaches, but unfortunately it's all I can do just keeping up with the present, which is quite a continual mess of its own, actually.

Does any of this make any sense? Probably not. So what else is new with me? Nothing, I guess.

Love from your Mother.
Love to Raymond from your Grandma.

PS I sincerely hope you haven't been forgetting to water the back garden!!!

'I think Grandma's going to be in the market for a really good deprogrammer one of these days,' Raymond said. He was picking at his fried egg with a fork. He seemed to be deciding whether to eat the egg, or place it under a microscope for further study.

'Don't worry about Grandma,' Sarah said. 'Instead, worry about what Grandma's going to do to both of us when she gets an eyeful of that back garden.'

It rained all the next morning. The Customer Service Center was humming with complaints about wet and torn newspapers, misdeliveries, and faulty billing statements. Around ten a.m. one of Sarah's floor supervisors began visiting the modules and signing clerks up for extended duty through the afternoon at time and a half.

'I can't, really,' Sarah told her. 'I'd really love to – I could use the money. But I've got to go home and drive my little boy to his treatment.'

Even before Sarah returned home that afternoon she was angry. She cursed every time the bus lurched, or she dropped her newspaper, or the sun flashed in her eyes. Her skin felt spotty and pale. She was afraid to check her reflection in mirrors.

When she arrived home she found Charlie's car in the driveway. Charlie and Raymond were playing checkers on the coffee table in the living room, sitting cross-legged on the carpet and sharing a sack of Mr Salty Pretzels.

Sarah said it over and over to herself as she marched up the driveway and unbolted the front door. The words had been building in her all day like steam in an engine. Now she knew where the words were going. It all made perfect sense; this was the moment she had been waiting for. And of course when she opened the door he looked at her like she was crazy or something, like she didn't know what was going on, like she couldn't make rational decisions for herself. All day long, every day of her entire life, that was the way men had looked at her. Like she was stupid or something.

'Sarah,' Raymond said. 'I had to call him. You forgot my appointment this morning. Don't be mad at Charlie, Sarah. Be mad at me. It's *my* fault. Charlie was just helping out.'

Charlie was saying No. Charlie was putting up his arms. Charlie was moving toward the door. Every ounce of energy she had in her body was for him. She reached for the ashtray on the bookcase. She reached for the cups and saucers from the tea tray. She reached for the tea tray. Of course she had forgotten Raymond's appointment. Of course it was *her* fault. It was always *her* fault things happened. It didn't matter what she did. It didn't matter what she did so she might as well do it.

Afterwards, Raymond cried on the living-room floor for more than an hour. Sarah opened a bottle of her mother's brandy and sat with it on the living-room couch, watching him, feeling the slow ebb of rage in her blood. It had been born in her blood, and now to her own blood it was returning. Sea to shore, sand to sea.

'I had a lousy day,' Sarah said after a while.

Raymond was lying face down on the carpet. He was crying so hard his shoulders bucked. His breathing was guttural and snotty.

Sarah poured herself another brandy.

'I could've used the money, that's all. I think about it all the time, sport. How much I don't have. How much I need to get. Everywhere I go I'm always late. There's other people there already. They're prettier than I am, smarter, they've got better boyfriends. I don't even know how angry I get sometimes. Sometimes I don't even know what my anger's about.' Sarah lay down on the sofa and rested the glass of brandy on her stomach.

'I keep making mistakes,' she said. 'Especially when I'm certain I'm right. And when I'm absolutely certain I'm right, I seem to make the worst mistakes of all.'

'It's totally hopeless, Sarah. It's totally, totally hopeless. You're a totally hopeless person, and I've got to stop worrying about you and start taking care of myself.'

Raymond was shoving his clothes into a cloth pillowcase. 'It's like living with a human roller-coaster, Sarah. It's like living with a human volcano. You do more damage to yourself than to others, but that doesn't mean I can take any more of it. Every man for himself, Sarah. Like Grandma said in her letter. If I can save myself, then I'll do what I can for you later. But until I *do* save myself, Sarah, then I'm a little too busy for your nutty hysterical nonsense.'

Raymond tied a knot with the loose ends of the pillowcase and tossed the lumpy sack into the bedroom doorway. Then he pulled open another bureau drawer and started filling another pillowcase. 'Charlie's my father, Sarah, and I need a pretty solid family support-unit right now. And maybe Charlie isn't a genuis, and maybe he *is* broke most of the time, but *I* like him, and if *you* don't like him that's *your* problem. I don't know if it ever crossed your mind, Sarah, and maybe Charlie isn't particularly intelligent, but you're not the biggest genius the world's got to offer, either. You are a very, very, I don't know, a *very*.' Raymond had stopped filling the pillowcase again. He was gazing across Sarah's right shoulder as if he were scouting for some spectral dirigible. 'Erratic, Sarah. I think I can say you definitely qualify as a very *erratic* person.'

Raymond was out of breath, his thin chest beating underneath his white cotton T-shirt like the chest of a small damaged sparrow. He swung the second pillowcase over one shoulder, then picked up the first bulging pillowcase in his arms as if it were an enormous stuffed toy he was permanently cross with.

'Now, Sarah, if you don't mind, I'm hauling my skinny butt out of here.'

Sarah was leaning against the wall, watching Raymond's quick industry as if it were some incomprehensible algebra.

'But you can't leave, Raymond. You're my son.'

'Screw that, Sarah. You can leave Charlie. Parents can leave kids. Then I sure as hell can leave you. I don't need it anymore, Sarah. I don't need living in somebody else's house in boring old south San Francisco, or leaving my school, or leaving Charlie, or dealing with weirdos like Grandma's friends every day of the week. In case you haven't noticed, Sarah, I happen to have a very serious blood disease, and before it gets too late I'm going somewhere to relax, which means, of course, my home, where I've got my own room and my own TV. And if you don't like it, Sarah, *stuff* it. Now, if I go across the bridge to the market can I catch the 27 to the city? Oh, hell. I've got a twenty Charlie gave me in my wallet. I'll call a cab if I have to.'

For a while after Raymond left Sarah stood in the bedroom doorway, watching an angle of moted sunlight lean through the window. She seemed to be drifting in the apex of a recession. The house was growing quieter and quieter, events were moving further and further away. Sarah sat down on the bed for a moment to catch her breath. The wallpaper in her old room still depicted little schoolgirls in frilly polka-dot dresses carrying their schoolbooks on frayed hemp ropes and being pursued by devoted, tangling puppies. After a while, Sarah heard the voice of the television reach up to her from the living room. While the rest of the world was receding, the television seemed to be growing louder and louder.

It's like living in a world full of strangers, Sarah thought, and went downstairs to the television. You live with these people, they give birth to you and you give birth to them. You sleep with them, and mix with their heat in rooms and hallways. But you never seem to understand them, and they never seem to understand you.

In the living room, the CBS *Evening News* had just started, with Morley Safer standing in for Dan Rather. Sarah stood behind her mother's stuffed chair and gazed at the TV. She felt herself beginning to drift into the white spaces of the walls and ceilings. Whenever she started to drift like this, even Charlie and Raymond screaming and wrestling on the floor couldn't shake her out of it. Eventually Charlie would lean over her while she drifted and give her shoulders a firm little tug, just to remind her what material existence was like.

'Hey there, Sarah,' he would say. 'Hey, baby. Where you off to, honey? Are you in a boat or a jet plane? Hey, Sarah. You want me to take care of dinner tonight?'

But now Charlie wasn't here. Raymond wasn't here. Her mother wasn't here. There wasn't anybody to remind her. If she wanted to remember, she had to do it all by herself.

On TV, Morley Safer was describing recent events in the Presidential campaign. There was one guy with a sort of lean gray face who kept pledging allegiance and talking about a kinder, gentler nation. There was another guy with a pudgy, distracted face who was riding around on top of a tank and wearing a silly helmet.

'Screw the both of you,' Sarah said, and dug into her handbag for her mother's car keys.

It was nearly dark. The lights were on in living-room windows along the street. Sarah turned into Skyline Boulevard and activated her windshield wipers as a thin mist developed. She could already see Raymond on the bus bench. He was braced between his two bulky pillowcases and reading something scrolled up in his hand. When Sarah pulled over to the curb she was thinking she wanted to say something, something slight, something that would hurt but wouldn't hurt too much. She thought about saying, What you and your father want. No matter how hard I try. Who cares about what I want, what I need. She rolled down the passenger window and leaned into the mist. Raymond didn't look up. He appeared pale and thin, his lips red and swollen. He wore an uncharacteristic vacancy around his eyes and mouth.

Sarah tried to think of something. She had to think of something soon or Raymond would continue drifting away into the soundless currents of himself. Further, further away, on little slips and counter-slips. She had to say something, but she couldn't think of anything. She couldn't think of anything to say, so she said the first words that entered her mind.

'Hey there, buster,' Sarah said. 'Say hey, sport.'

Causing Raymond to look up from his *X-Men* comic book.

THE OTHER MAN

'It's a chemical thing,' Edward Thomas said, prone on padded leather, cleaning his nails with a twisted paper match. 'There's no use telling me whether he was *really* there or not. That's pretty inconsequential. What *is* consequential is that whenever I enter that house I *know* he's been there. He has just fucked my wife, eaten food out of my refrigerator and sat on my toilet. I'm not talking about breach of propriety, you know. I'm talking about a violation that's practically *cellular*, for chrissake. There I am, home from work, placing bread and milk on the bookshelf beside the front door and taking off my coat, and I *know* it happened right there, right there on the living-room floor. He mounted her from behind while she jerked her head wildly from side to side. I've never seen her make such a fuss before. I never even thought she *liked* sex that much, if you want to know the truth. And then the noise they're making. This isn't my wife; it's some goddamn truck-stop waitress. There's a moment in which this imagined infidelity actually *invests* me, like a murmur of the stomach, or a pulse of the blood. It's a sort of biological event; my entire system responds. I'm breathing quickly, my temperature rises. It's an anger that pulls at my stomach – an anger not at my wife, you understand, whom I love, but at this vision itself, this glimpse of some ungraspable life. I don't know if I can explain it better than that. I want to reach into my stomach and haul it out, I want to pull the memory of that vision out of the fabric of my cells and tissue. I have seen something real which is not real. I'm willing to accept that, you see. I'm perfectly willing to continue believing that my wife, Rachel, is a faithful woman who loves me, but that doesn't prevent another physiological world from moving in my blood and beating in my breath. It affects me as strongly, you see, as if these things had actually happened. I guess I really don't know how to explain it. I can't sleep any more. I wake up with these awful anxiety attacks, my heart pounding. I've tried aspirin and Valium. I started sleeping with Phil Brady's wife a few weeks ago, just for a sense of vital retribution. I bought a personal computer. Sometimes in the middle of the night I drive to Ventura, Escondido, even Santa Barbara. I can't keep my mind off it. I'm really hoping you can help.'

Dr Tobias's wristwatch chimed twice, and Edward craned his neck around. The patent leather creaked. From this angle, Edward could glimpse only Dr Tobias's very casual cotton shoes.

'My policy with new patients is to adopt a very cautious attitude the first few weeks,' Dr Tobias said, shaking Edward's hand again before he left. 'If you have any trouble, feel free to call, and I'll see you again next Tuesday at one o'clock.'

'He's a strict Freudian,' Rachel said, offering Carla Sara Lee Danish in its aluminum tray. She licked her fingers. 'That means he's going to delve into Edward's subconscious. Edward's going to see him three days each week, this week and next. Then he sees him twice a week for two years. Edward lies on the couch and Dr Tobias sits behind him. Edward's not supposed to look at him, or else he'll suffer transference.'

'Is he good looking?' Carla sat on the edge of the sofa with her legs crossed. She was wearing a short white skirt which looked good on her.

'Dr Tobias?'

Carla bit her Danish and said 'Mm-hmm,' catching a sudden flurry of crumbs with her hand.

'Yes,' Rachel said. 'I guess so. But I think he's married.'

'Yesterday I found a can of his shaving cream. It was in my bathroom, on my sink. Edge Shaving Gel for Sensitive Skin. I never used Edge Shaving Gel in my entire life. I use Gillette Foamy.' As usual, Dr Tobias sat so quietly Edward could only hear him breathing. Sometimes he imagined that as he talked, he was actually trying to reel Dr

Tobias in on the line of his voice. He could feel Dr Tobias's presence that way – like some weight hidden beneath water. 'I could have brought it along and shown it to you, but I didn't see the point, really. I mean, I don't want this all to degenerate into some pointless argument about whether I'm deluding or not. I know ultimately I'm doing nothing else but. It's all one long dream, you know. The other man, my wife, my therapy. I take all that for granted; I'm not what you'd call naïve. The can of Edge Shaving Gel is perfectly real; the other man may very well not be. One way or another, I still dream about him. In my dreams, he and I get along famously. He never makes me feel threatened; he always makes me feel as if this whole thing is quite a bit of good luck on both our parts. It's not him that we're trying to figure out anyway, is it, Doctor? It's me, and the world as I've fashioned it. We're drawing a sort of map. This is where Edward's addled brain ducks, turns, pirouettes, swerves. These are the objects he sees in his way. I guess I should've gone looking for one of those existential psychologists – is that what they're called? I mean, a strict Freudian. Really. But that was a sort of existential choice itself, you see, because I believe in the family romance. I really do. Mother, father, daughter, son. Dark dreams of incest, immolation, taboo. Those are the dreams we need, I think; they make a family *work*. Otherwise, what have you got? Happiness, fidelity, more happiness? Everybody being happy happy happy all the time? That would be a little hard to take, wouldn't it? Is that the kind of world you want your children to grow up in?'

Behind him and out of sight, Dr Tobias lightly, audibly belched.

Sometimes at night Edward even dreamed of the other man. They ascended strange buildings together in escalators and elevators. They were always on their way to some imminent cocktail party to which Edward's wife, Rachel, had also been invited but was not really expected to arrive.

'Eventually they become great buddies,' Edward Thomas said, flicking a Tareyton ash in the glass ashtray balanced atop his chest. 'They go everywhere together. Movies, fishing, sporting events, RV shows. Neither has ever fully enjoyed the pleasures of male companionship before. There's something regenerative about this as opposed to a mere sexual relationship. Neither feels *drained*. Neither tries to bring the other *down*. There's absolutely no sense of competition between them. This doesn't mean they don't appreciate or enjoy the company of women. Not by a long shot.' Edward crushed out the cigarette, turned to replace the ashtray on the table beside his bed. He heard the bedroom door open, brushing against the shag carpet.

'Edward?'

Rachel's face looked in from the dim hallway. The bright bedroom still swirled heavily with smoke like some domestic cabaret.

'Yes, honey?' Edward fished another Tareyton from his shirt pocket.

'Are you all right in here?'

'Just thinking out loud is all.'

'Oh.' Rachel's eyes flicked about the room a few times. 'Dinner's ready,' she said finally, and pulled the door slowly shut again.

'I don't understand why you don't go out with somebody else, then. I mean, if he *thinks* you're out balling some guy, you might as well have a good time, right? That's what I say.'

'You don't understand, Carla.' Rachel put down her menu, looking over her shoulder at the salad bar. 'Ed doesn't think I'm *cheating* on him.'

'I don't understand, then. I *really* don't understand.'

'Ed's *ill*, Carla. I've explained this all to you a hundred times. He has delusions. He *knows* they're delusions. He needs a lot of love and patience right now, and I'm trying to give it to him.' Rachel opened her purse and pulled out a long, crumpled Kleenex. 'It's not a matter of he doesn't *trust* me. That's what you don't seem to be hearing.' Examining it first carefully, Rachel placed the Kleenex against her nose and blew.

'Okay, maybe I don't understand your crazy husband, but I know what I'm hearing, girl.' Carla snapped her compact shut and pushed it abruptly into her purse. She took her smoldering cigarette from the glass ashtray. 'Your husband says you're screwing some guy. He says, listen to me. Listen to all my crazy stories. Sit up all night, sit around all day, listening to my crazy stories about how you're screwing some guy when you're *not* screwing some guy—'

'Can I take your order?' The waitress held out her order pad and pencil. She looked about seventeen.

'Just a second, baby.' Carla leaned earnestly across the table, stage-whispering. 'Because now *you've* got to start listening, Rachel. Now *you've* got to start reading the writing on the wall. What do you think you're going to do? Spend the rest of your life wondering what's best for *Edward*?'

'I'll come back,' the waitress said.

'No, of course not. Because you know why? You're not stupid, that's why. So what do you do? You've got Ed thinking about Ed, and you've got you thinking about Ed. That makes two people thinking about Ed. And where does that leave you, Rachel? I'll tell you. It leaves you nowhere, it leaves you with nobody. It leaves you nowhere, with nobody, doing nothing. You think *I'm* going to worry about you, girl, if this loopy husband of yours fucks you over? Then you've got another think coming. I'm looking out for me, girl, because I'm like Ed. That's where Ed and I are a lot alike.'

Rachel felt slightly disoriented and fuzzy, as if she had been slapped. She had never heard Carla talk to her like this before; her face felt hot and red. She got up and went to the bathroom and examined herself in the mirror; then, briefly, she cried spontaneously for almost a minute, without any reason at all. They ate their side salads and finished their drinks in silence. Afterward, when she offered Carla a ride home, Carla just snapped at her again. 'I'll walk,' she said, and that was exactly what she did. Rachel couldn't understand why Carla was in such a terrible mood. After all, she thought, *I'm* the one that should be mad. I think I was being awfully patient with her.

But even later, after she arrived home, she felt more ashamed than angry. The parameters of the entire house seemed to shift as she put down her bag and sat on the living-room sofa. It wasn't as if Carla knew anything. After all, how many good relationships had Carla ever had? It's easy to say just think about yourself when you live alone, Rachel thought. When you've never had the courage to maintain any type of commitment with anyone but your own self. Outside, beyond the living room's uncurtained picture window, the sky grew suddenly dark, and Rachel wondered if it might rain. The air had been heavy all afternoon.

The house was very quiet now, and Rachel lay down on her side on the sofa, the remote control gripped loosely by her right hand, *The New Hollywood Squares* playing soundlessly on the television. Rachel wondered idly whatever happened to the old Hollywood Squares. Nothing significant came to her mind for quite a while. She felt more vacant and muted than sad. Then, eventually, motionless before the television's abstract flickering, she wondered what the other man was really like, and whether he made love to her in Edward's fantasies with great passion or calm assurance. Edward had described him to her many times. He wasn't handsome, really; he was quite 'normal' looking, with brown hair and dimly blue eyes. Usually he wore Levi's and shirts with button-down collars. He was boyish and energetic. He wasn't like other men. His suggestions were always relaxed and disinterested, and always made a lot of good common sense. She didn't have to explain to him more than once that she loved her husband, that she would never do anything to hurt him. When he kissed her mouth his hand moved confidently around her waist. He pulled her close. His right leg moved between hers. Gently he lowered her to the floor, his left hand brushing the backs of her legs. 'I love you,' he said, but it didn't matter, didn't matter what he said. She took a sudden breath when he entered her, training her eyes upon the cut crystal and silver tea set above her in the glass cabinet. 'I love you,' he said, 'I love you, love you,' pressing her against the rough carpet. 'I love you,' he said, but she only loved her husband, she told him, but don't stop anyway, love me anyway, keep loving me even when you hear his feet on the stairs, even when you hear his hand on the door. The entire floor seemed to contract suddenly; the entire world seemed to swirl. And then she was coming in his arms, and the door was opening, and Edward with his briefcase was staring down at them, not surprised but fulfilled, always expecting it, always anticipating the thrill of the moment when she came, breathless, swelling against the rough carpet, crying, You, you, baby, now, in the hard, enduring embrace of this other man.

When Edward arrived home around seven-thirty, the house was empty. He wondered if this was the very silence he had been somehow expecting. There was something familiar about it already as he lay his briefcase on the bookshelf and stepped into the living room. He thought motionlessly for a moment, listening. 'I stepped into the house,' he said out loud. His voice was gentle, as if he were trying to reassure someone. 'I had just gotten home.' He stepped into the immaculate living room and saw the Pledge-bright woodwork and furniture, the blinking digital clock on the video recorder. 'Suddenly all the voices in my head went quiet. I tried to remember what the other man looked and sounded like, but I couldn't.' He stepped into the hallway, the kitchen, the bath. 'I went into the kitchen. Everything was clean. Even the dish towels had been recently washed and folded. I felt strangely hypersensitive, predatory, keen. I wouldn't miss a sound, a motion. But there weren't any sounds. There weren't any motions. I went into the bedroom, talking out loud to myself in the empty house, trying to imagine an object of my conversation, but unable to visualize anybody, feeling only the secure silence situated somewhere in the house like an individual presence. I went into the bedroom, the hallway again, the living room. The television, VHS, sofa, bookshelf. I was trying to track the absence of something, I was trying to locate the place where it had ceased to be. It was like a racial memory, like the dream of falling from high trees. I knew everything would be all right in the long run. I knew everything would work out for the best.'

A few weeks later he received Rachel's letter, posted from Fort Lauderdale.

> Dear Ed. I really can't explain. I've tried to think of a way I could explain, but I can't explain. I still think about you all the time. Please don't hate me all the time.
>
> Love, Rachel.

Sometimes it woke him in the middle of the night. Other times, in order to delay going home to an empty house, he ate dinners at restaurants or attended double features at the cinema mall. The only part of the day he actually preferred, however, was his time at work, where he tried to arrive especially early each morning. The janitors were usually still pulling large gray dustbins on trolleys down the polished hallways, or operating roaring, man-size vacuum cleaners with corrugated, squidlike rubber hoses. Eventually, however, he had to return home and face it. He didn't feel frightened so much as deeply excluded. He felt hurt; he felt inarticulately sad. He wanted to be readmitted to that silence, that absence. He wanted to hear the language of that darkness again.

Three times a week he sat on Dr Tobias's couch and stared at the ceiling's white acoustical tile. He never felt like saying anything; he couldn't think of anything to say. He tried to imagine Rachel on jets, visiting impossible countries, hearing the other man's voice in her head. What did he tell her? Did he convey advice, affection, information? Edward couldn't imagine. It was as if he couldn't comprehend language anymore. Strangely enough, it was when Edward grew silent and desultory that he somehow elicited the voices of others.

'What did you do today?' Dr Tobias asked.

'I went to work.'

'Did you speak with anybody?'

'My secretary. I made some phone calls.'

'Have you made any plans for tonight?'

'I'll probably watch television.'

It was as if Edward could comprehend only literal, dry events. After only a few weeks or so his silence even began to infect Dr Tobias, who would sit alone in his office for hours after Edward's sessions and gaze silently out the window at high converging power lines and the flashing blue and green lights of planes descending over the airport. When he went home, Dr Tobias's wife would serve him his favorite dishes and sit silently by his side while he stared absently at the evening paper or smoked his pipe and stirred the ashes in the ashtray with his charred and twisted pipe cleaner. Sometimes Mrs Tobias would go to bed alone, worried by her own vague suppositions, while Dr Tobias sat awake late into the night, not thinking of anything in particular, listening to Brahms's *Requiem* on the compact disc player and feeling an odd, indefinable melancholia suffusing his blood like a drink of alcohol.

It wasn't until the following winter that Edward began to show signs of 'progress.' A few days before the winter solstice he arrived and took his customary seat on the sofa. The weather had been unusually chill for Orange County, and some forecasters were even predicting snow. 'Sometimes you don't know why,' Edward said. Deep underneath the floor the massive industrial heaters started and began to blow the office full of warm, abrasive air. 'It's just something you feel, not something you can talk about. It's not chemical, it's not your parents, it's not even the newspaper. It's some sort of invisible condition. It's that part of the world you can't explain.' For the remaining hour Edward lay silent on the sofa, only speaking in order to answer a few of Dr Tobias's cool, aimless questions.

On his way home that evening he was caught in a traffic jam on Irvine Boulevard. The uncustomarily cold weather seemed to elicit a rather sudden, edged anxiety from everyone. Edward looked in at other cars where drivers gripped their wheels angrily or gave sudden fingers in the Pacific's general direction. Sometimes long white limousines glided smoothly past, their engines generating warm, humming vacuums of space as well as sound, their occupants' faces occluded by dark, glimmering windows. At home Edward knew he would be able to relax. He had learned, and was learning, not to let things bother him so much anymore. He still had his home, his job, his health. In the distance there was a crack of sudden thunder, and the entire atmosphere seemed to give a little. It might rain at any moment, and Edward imagined himself at home already, sitting in his living room and watching television, the other man's discarded possessions heaped on the coffee table before him. The Edge Shaving Gel, Trimco nail clippers, flexible black plastic comb. A linty button and some stray change Edward had fished from behind the sofa cushions. A Casa Maria matchbook with an unlabeled phone number printed inside with a blue felt-tipped pen. If he didn't try too hard to make sense of things, Edward knew that things often made sense of themselves.

Edward never saw his face or heard his voice again, and eventually he even stopped speculating about the other man's name.

THE PROMISE

Nothing is accidental except physics.
JOHN BERRYMAN

Ever since she could remember, Amanda Hoffman had admired intellectual men. Men like her third-grade teacher, Mr Fogerty, who drew complex diagrams of microscopic organisms and planetary cycles on the squeaky blackboard. Or men like James Herriot, who had written Amanda's favorite novel, *All Things Bright and Beautiful*, concerning a veterinarian in England who was educated enough to treat the ailments of pets and livestock, and compassionate enough to appreciate the complex psychologies of the people who loved them. Intelligence gave men a certain abstract depth and volume, Amanda thought. Intelligent men took up more space than ordinary men, embraced vaster geometries and dimensions. Their homes contained the books of great writers, the framed museum prints of brilliant sculptors and painters, and the bright web-like harpsichords of world-renowned musical geniuses like Mozart and Bach.

When Amanda was seventeen she married her composition instructor at Long Beach City College. His name was Dr Wayne Stanhope, and together they rented a small one-bedroom garden cottage on Ximeno Avenue in Belmont Heights. They purchased a cream-colored living-room set with convertible sofa, and Wayne installed his private den in the compact rear bedroom. Tall unvarnished pine bookshelves, a wide black particle-board desk with matching printer-stand and filing cabinet, a bulletin board studded with bookstore and college promotion leaflets, a calendar, an answering machine, and a clock. Every weekday morning Wayne drove to the college where he taught comp, remedial comp, advanced remedial comp, and a special section of 195A – an instructor-designed honors course in something Wayne called the Literary Psychology of Social Movements.

After Wayne left each morning, Amanda liked to take her coffee into Wayne's empty den and sit at his wide impressive desk. She felt like one of those children in the fairy stories who enter the forbidden rooms of magi and sorcerers to discover strange new continents and oceans, creaturely transformations and stunning, intricate riddles. All the time she sat there, Amanda couldn't stop smiling. Here were books on archetypes and ids, rhyming patterns and sonnet schemes. Here were books on modernism, deconstruction and New Age philosophy. Feminism and astrology, Gurdjieff and Jung, Arthur Janov and Wallace Stevens. Wayne kept his crabbed black handwritten notes for a book on Pound in a stack of stiff yellow legal notebooks, bound together by one flat red rubber band. Amanda liked to hold the weight of those notebooks in her lap while she drank her black coffee. This was Wayne's work. This was the way Wayne felt when he was alone at night working in his den.

Amanda worked six afternoons each week as a secretarial temp for Gals Friday in the Redevelopment District, typing, filing, and disregarding the stern, often humorless advances of large flushed men in polyester leisure suits, men with soft red hands, receding hairlines and minty white teeth. At lunch Amanda would eat leftover salads from opaque plastic Tupperware, cookies and sandwiches, quiches and crudités. At the end of each week she promptly presented her payroll check to Wayne, who always perused it with a sour, condescending scowl before he folded and placed it without further comment in his faded brown leather wallet. 'We don't like to think about money any more than we have to,' Amanda told her friends at work. 'We have Wayne's car, and a nice house, and even free use of the yard whenever we want to. It's not like we need a lot of money and things like that to be happy. It's not like Wayne and I need all those shallow, materialistic things everybody *tells* you that you need to be happy.' On Saturday mornings Amanda liked to stop at the Valley Street Thrift Shop on her way to work. The Valley Street branch was the best Thrift Shop in Southern California, where you could always find quality brand-name items that rarely cost more than a dollar or a dollar seventy-five.

Every evening Amanda returned home in time to prepare coffee and nutritious, well-balanced meals for Dr Wayne Stanhope, who was the sort of deeply spiritual, philosophical man who might very well neglect his diet and physical well-being altogether if Amanda didn't keep a careful eye on him. Fresh fruit and vegetable salads, home-made soups and vegetarian pastas, fresh-squeezed juices and cracked whole-wheat breads and rolls. Wayne always looked immaculate and imperiously wise when he returned home from work. He wore fine white cotton shirts and pressed corduroy slacks. His shirt sleeves were casually rolled back, his knit ties loosened. His intelligence was like warmth or radiation. It was geothermal, and radiated in all directions at once, much like the unified field theory that occupied Wayne's thoughts constantly.

'The unified field theory,' Wayne said calmly, his hands clasped on the tiny oak table Amanda had purchased at a neighborhood garage sale, 'is the theory Albert Einstein was working very hard to complete when he died – quite mysteriously, I might add, during a time when the Soviet Union was exercising its iron red grip across Eastern Europe. Like most great intellectuals, Einstein's mind was concerned with the big ideas, and not all those silly, petty ideas that occupy the minds of most people, like television or politics.' Wayne intently examined the dark grains at the bottom of his coffee cup, like a biochemist with a petri dish. Amanda had purchased their matching coffee mugs at the Lakewood Mall for their first month anniversary. Wayne's mug said MR DR, and Amanda's said MRS DR.

'Would you like some ice cream?' Amanda asked, beginning to clear the dishes. 'I bought some vanilla ice cream at the store today.'

Wayne put down his coffee mug and gazed up at her. Sometimes, Amanda couldn't even gauge the depth of Wayne's eyes. All she could detect were the concave, distorted reflections of bookshelves and windows.

'I don't really like vanilla ice cream,' Wayne said softly, with neither anger nor remorse, the same way he read poetry aloud from a book. It was as if his words were dead laminated fish being placed on the table one at a time, with glassy black eyes and thick, cartilaginous spines. 'In fact, I don't ever remember telling you I liked vanilla ice cream even once in all the time I've known you.'

*

Weeks before her pregnancy was verified by the obstetrician, Amanda could feel consciousness stirring down there. Formal, quiet, potent, superior, hard. It possessed a deep metrical clarity, like machinery or poems. It assembled itself from stray atoms, molecules, enzymes, cells. It gathered minerals and properties, random bits of matter and implicit propositions. It wasn't even discrete yet, but already it seemed to have its own cool ulterior purposes very firmly in mind. It was partly Wayne's and partly Amanda's, and partly something that was neither Wayne's nor Amanda's. It was as if some deep part of Amanda's own body were betraying a secret, molar race of echoing blips and beeps, as incomprehensible as the philosophies hidden inside Wayne's books. Amanda felt sick every morning when she got out of bed, and every evening when she returned home from work. She felt thin, enervated, listless and pale. Her teeth and gums ached. She felt very happy and very content. This was another world down here. This was another planet entirely, filled with different landscapes and different trees.

'I remember when I was pregnant with you,' Amanda's mother told her one afternoon over Oreos and instant coffee. Amanda's mother's name was Esther Rockweiler, and Esther lived in Fountain Valley in the top floor of a modular concrete four-plex she had purchased shortly after divorcing Amanda's father in 1975. 'I threw up every morning and every night. I'd go out for a nice meal or lunch, you know, and then I'd spend the next half hour or so in the ladies' room throwing up. And when I say throw up, I'm talking severe vomiting. I thought I was going to die. I thought I was going to throw up every organ in my body eventually. Kidneys, gallstones, pelvis, you name it, the works. I was throwing up entire mountains and cities, Volkswagens and office furniture. And then when you were born – Jesus. They had to sew me up afterwards with piano wire. I think it was some sort of fly-by-night low-cost medical insurance your worthless father dug up somewhere. They cut me straight up the middle with old gardening shears or something. I'm telling you, they could have used me afterwards for bookends. Having a baby, well.'

Esther lit one of her Newports and took a long, rustly drag. She gazed out the bright window at Amanda's landlord, Mr Anderson, who was spraying the gardenia and rose bushes with insecticide from a large, rusty steel drum. 'All I can say, honey, is that anybody who wants to have a baby should probably get his or her head thoroughly examined. Me, I finally decided abortions were a safer bet. But if a baby's what you really want, then there's no use me trying to tell you otherwise. I'm afraid babies happen to be one of those things every woman's got to figure out for herself.'

'When you are born, you will have everything in the world you could ever desire,' Amanda whispered at night, often hours after Wayne had fallen asleep. Sometimes Wayne was still wearing his glasses, a stained library book propped open in his lap, coffee cooling in his mug on the end table. 'You will have your father's brains, and your mother's good looks, even if I do say so myself. No matter what you do or where you go, you will always have a home to count on, and parents that love you. Your name will be Robert if you are a boy, and Jacqueline if you are a girl. You are coming from a pretty good gene pool, by the way, except for a small part of my side of the family, where most of the adults are totally crazy all the time. Wayne's intelligence genes should make up for all that, though. Your father is a brilliant Instructor at Long Beach City College, and everyone who ever meets him always says that he definitely has his head together.'

Often, Amanda sat up in bed while she spoke to her child. Attached to the splintering headboard, Wayne's clip-on bedside lamp cast long shadows across the bedspread and floor. Wayne snored slightly, muttering, turning. While Amanda rested her fingertips on her resonant stomach, she thought about how small the planet was, how intimate things like nations, moons and asteroids were. At times she swore she could even hear the beep and swerve of golden satellites in the sky, but whenever she told this to Wayne the next morning at breakfast, he always told her she was being nutty again, just like her crazy mother.

'The reason I am whispering now, is because this is what is called hothousing.' Sometimes Amanda stroked her warm, expanded stomach when she spoke. Sometimes she tapped it with one gentle inquisitive finger, as if she were keying a particular command over and over again into a word-processor. 'Hothousing means the educating of a young baby before he or she is even born. That's why I bought the French language tapes today. That's why I play a lot of symphonies now, a lot of operas too. Because I want you to have a head start on all the other children in the world.'

*

Amanda felt thrilled and dizzy whenever she contemplated the accumulating presence of her first child. She began browsing through Wayne's old university textbooks, looking at diagrams of hereditary material, meiosis, fertilization, the first graph-like buds of the neural fold, ectoderm, and neural plate. It wasn't merely a body, Amanda decided; it was a sort of process, more abstract than skin and muscle; spatial, fugitive, errant and vast. Amanda realized her blossoming child was approaching her across wider, more theoretical landscapes than those of her own body. Her child was like astronomy. Her child was like math. Her child was like the secret emanations of plants. Over succeeding days and weeks, Amanda grew increasingly convinced that extraordinary signs and patterns were emerging from the world.

Often when Amanda returned home in the afternoons she could hear it beating, beating. The swirling eccentric motes and streaming sunlight of it, the hanging ferns and shady brick porch. For a moment, she lacked any sense of direction at all, as if even the notion of distance had collapsed around her. Everything in her apartment felt like an impediment, abrasive, dissonant, edged. Her nerves seemed strung together by a fine, sharp filament of thread, contracted and glistening like the net of a spider. The real world was somewhere else, she thought. The real language, the real place, the real beat of the real heart. Everything else just got in your way, prevented you from grasping finer realities than them. The sofa, glimmering hardwood floors, warped aluminum kitchen cabinet and fixtures, gauzy sun-stained curtains and high, unavoidable bookshelves. Everything solid and permanent was like a contradiction. It just confined you to this hard apartment which wasn't really real at all.

Amanda sprayed the cottony, sunlit rooms of her miniature house with home deodorizers and air fresheners. She dusted the leaves of houseplants, the glistening Brodart spines of hardback books, the orderly and neatly appointed lamps and countertops and clocks. She treated the rich, dry woods of tables and chairs with conditioners and polishes, swept deep into the coarse fibers of carpets and upholsteries with a stiff broom, and filled a plastic green bucket with Pine-Sol and steamy water. Then she irrigated a dry gray sponge and began in the bathroom. She scrubbed the white tile and sandy gray connective mortar, the sink and toilet, the clogged shower grating and dusty, over-painted window-sills. She wiped rusty oxidation and mould from coppery colonic pipes and misty refrigerator cabinets.

As Amanda reached beneath the surface of her home, she felt herself growing cleaner and more pure, as if she were reaching into the heart of her own body where everything was perfect, germinal and intact. 'The world is a very big, confusing place,' Amanda said as she worked, mopping the floors with astringents and cleansers. She liked the high hard scent of them, the scent of old churches and brand-new bathrooms. 'But that doesn't mean there aren't simpler places we can go. It's a little like geology, like digging into history with a shovel. Mesozoic, Paleozoic, and whatever that other one was, the one with the dinosaurs. Dig down deep enough and eventually you come to this, this hard primal floor of the world, this place that existed long before history ever got here.' She stacked clean dishes in wooden cabinets. She polished silverware. She repaired shirts and curtains and drapes. She washed clothes and towels and linen and ironed them. During the imminence of her first child, Amanda believed she could live in both the future and the past at the same time.

Then, one afternoon in early August while she was filing precedents for a Long Beach public relations' attorney, Amanda suffered a miscarriage, and the entire world as she knew it changed irrevocably into something she did not understand.

'It was never even alive to begin with,' Wayne said, seated strangely beside her hard white bed with a distant, slightly perturbed expression, as if he had just returned from some pointless and exasperating errand somewhere. 'It was just a bundle of tissue, impacted hair and marrow, waste products, food. It was as if your body started sending out the wrong messages, honey, and your metabolism never got its act together. It was just muddled language, random, piano-like, untranslatable. Funnily enough, I was talking to my class about the very same thing just this morning. It's practically the same paradigm-formation syndrome which afflicts all linguistic and cultural orientation systems in our world today. I wouldn't worry too much about it, though. Your doctor said we can always try again.'

Wayne took the hard vinyl chair beside Amanda's bed. In his lap he opened a heavily warped and underlined paperback copy of Karl Popper's *The Open Society and Its Enemies*. 'Oh, by the way,' he added, turning a page. He pulled ruminatively at his lower lip. 'Your mother was here about an hour ago while you were asleep. But then she left.'

Suddenly everything about Amanda's life seemed perilous and indistinct. Meals, bed, bathing, sunlight, houseplants, kitchen utensils, appliances. Sometimes she just sat and watched Wayne sitting across the living room from her. He was reading a book, and she was still wearing her bathrobe. While Wayne read, he pried at his ear with his little finger, or scratched his thighs. He was sitting on the gray osier recliner. He was drinking hot tea and reading the paper.

'It doesn't even feel like my own body anymore,' Amanda said one evening. She didn't realize she had started to cry. 'I feel so terribly emotional all the time, and I don't know why. I don't even know what I'm crying about, or why I feel so sad.' All day Amanda sat in the same blue robe with the torn blue pockets she had worn home from the hospital. 'All I ever do anymore is sit here and watch television. Did you know, Wayne, that there's not a single good program worth watching on TV all day long, even when you've got Cable?'

Wayne was levering the lid off a can of Dennison's Chili Con Carne with a butter knife. Then the chili disengaged from the overturned can with a flat, wet sucking noise, and plopped into a sizzling saucepan. 'Your body's undergoing a lot of really heavy hormonal changes right now, doll-face,' Wayne said. 'All your emotional and linguistic communications lines have gotten themselves cut, tangled and crossed. There are certain large fundamental spaces in your body and brain that still think you're going to have that baby. In some ways, you're a perfect living example of what Jean Paul Sartre often refers to in his writings as the *pour soi* – or is that the *en soi*? I'm always getting those two confused. How about if I fix you a glass of red wine, honey? How does a nice glass of red wine sound to you?'

Wayne was holding an upraised, chili-stained tablespoon in one hand. Amanda looked past him at the piles of dirty dishes in the sink, the stray drifts of feathery dust on the floor. The large green Hefty bags were overflowing with empty milk cartons, frozen food containers and crumpled paper towels. The house kept filling up with this stuff, this stuff, Amanda thought. Suddenly it was arriving from everywhere at once – supermarkets, post office, stratosphere and garden – obscuring and retexturing things, covering the entire world with a flat gray integument of dust. Soon the entire house would be gray and thick with it.

'I don't want any wine,' Amanda said. She touched her face with the palms of her hands. The tears on her cheeks had begun to dry. 'I just want to feel better. I just want to be myself again. I just want to go back to work and see my friends.'

Amanda did not begin to feel better over succeeding days and weeks. Instead she felt nothing at all, a long beatless routine of buzz and irreproach. Alone all day in the tiny guest house, Amanda felt like a passenger secluded beneath the hull of a compact, rudderless boat. The entire world seemed to be moving secretly around her; other times, she didn't care if the world was taking her anywhere specific or not. She never fixed herself sensible meals anymore, and nibbled fugitively at broken cookies and stale crackers from the bottoms of neglected packages and boxes. Usually she sat in the living room and gazed out the slatted venetian blinds at the world's bright mirage: white sun, green leaves, birds, bees and telephone poles. As far as Amanda was concerned, the world simply wasn't worth knowing anymore.

'It's a beautiful day, doll-face. Would you like to sit outside in the garden with me? Would you like a little sip of my iced tea?' Wayne was wearing a wide-brimmed straw sun hat, Levi cut-offs and thin plastic flip-flops. He did not look especially professorial or intelligent, Amanda thought. He wasn't even wearing his glasses.

'I don't want to go outside,' Amanda said evenly. Now, she thought, *I'm* the one who must explain things to Wayne. 'Sometimes I just like to sit by myself and feel sad if I want to. And that doesn't mean there's anything wrong with me, either.'

'Hey, baby. Pax, sweetheart. My heart's just not into all this *J'accuse* you've been firing my way lately.' Wayne raised his hands in a gesture of facetious submission, as if he were confronted by train robbers in an amusement park. 'I wasn't saying there was anything *wrong* with you or anything, honey. I just thought you might like a little sun, that's all. So don't bite *my* head off, Mandy. Don't try to make *me* feel like *I'm* the culprit here.' Wayne poured himself another tall, rudely abrupt glass of iced tea. Then he took it and his latest issue of *PMLA* outside to the green plastic sun chair and glistened alone, pale and miraculous-looking, in the bright hot sun.

'I'm not trying to make Wayne feel unloved,' Amanda told their marriage counselor, a fortyish Asian man named Dr Wong who used to be a dentist. These days, however, Dr Wong referred to himself as a Marital Interaction Coordinator. 'I know I'm being selfish. I know I'm having trouble exhibiting affection. I'm sure I still love Wayne, but it's just getting harder to be around him anymore. He's always trying to tell me what's wrong with me. He's always calling me unpleasant things, like negative-minded, or hormonally maladept. But I don't care how critically minded you both get about me, Dr Wong. I don't think there's anything wrong with me, and I refuse to be put on any sort of medication. My father was on medication when I was little and he was always throwing things and screaming at my mother. I know I'm sad, Dr Wong. But that doesn't mean I'm unhappy.'

Wayne was growing broody and restless, and often tried to touch her at night while she slept. Eventually Amanda began wearing a thick cotton night-dress, fully buttoned up to her neck. Some nights Wayne abruptly got out of bed and went into the living room. He made a lot of rough breathy noises. He muttered angry things he wanted her to hear, flung pillows about and thumped them, eventually fell asleep on the couch. It always helped Amanda relax those nights when Wayne slept alone on the couch.

Then, on a Saturday afternoon in April, Amanda ventured outside to mail a birthday card to her Aunt Betty, and discovered the world was filled with children. Children on bikes, and children grinding intrepidly away at pavements with wide plastic Hot Wheels. Children with dogs and children with bright multi-colored balloons. Children running and crying, hiding and swearing. Some of the children loitered together on street-corners, huddling against the wind to light frayed cigarettes purloined from their mothers' purses. They showed each other comic books, snails, pieces of candy. They hit each other with rocks and sticks. Sometimes they just sat on the curb and stared into outer space, picking absently at the broken parts of their shoes, as if the afternoon would last forever.

Amanda turned and walked back up the garden path to her house. She went inside and left the front door open. She sat down on the sofa. Distantly, she realized she was still holding Aunt Betty's birthday card in her hands, votive, angular and white. The sun flashed through the multiple lenses of leaves, and Amanda could still hear the sounds of children playing, sourceless and proud, like the rush and organized confusion of waves or communities of insects. It was as if she had drawn the sound with her, like a rough hempen net filled with glistening silver fish. The sound was moving into the house now, the garden, the floor. It was beginning to agitate and pull at the deepest assumptions of things: the grass, the roots of old trees, the splintering fence posts, the deep concrete foundations of large and undeniable houses.

'You should count your blessings – that's what *I* say,' Esther told Amanda, over and over again while Amanda cleaned every inch of the house. Esther was chain-smoking some new cigarettes that came in a bright yellow package with lightning bolts that loudly proclaimed them THRILL! A NEW ADVENTURE IN SMOKING. 'Now you can do like I've always told you,' Esther said, while Amanda crawled around the kitchen linoleum on her chafed knees, reaching underneath the stove and cabinets with soapy sponges and harsh solvents. 'Now you can try getting *your* life together, Mandy, before you go pissing it all away over some goddamn baby. Like I've tried to tell you, have a baby and your life's not your life anymore. All day you're stuck with it, feeding and washing, picking it up when it cries, just waiting for it to be sick again. And just how long do you think Wayne's going to stick around, Mandy, once you start adding a little weight around the middle, and you're tired all the time, and never think about anything but sleep? Wayne's a nice young man, Mandy, and he has very gentlemanly habits and all, but that doesn't mean he's stupid.'

Amanda hauled all the fat, heavy bags of trash outside to the alley. She swept up long intricate cords of dust, washed dishes and pans, scrubbed floors and upholstery and blinds. After Esther left with her cigarettes, she opened all the windows, washed and ironed all the drapes. She wore white plastic dish-washing gloves, and tried to think nothing but pure white thoughts. Repeatedly her large yellow sponges absorbed flat black whorled layers of dirt, and she rinsed them in the sink. Then she washed the sink and sponges with ammonia and bleach. Before she knew it, Amanda found herself developing a secret affinity for water.

Wayne increasingly granted her less and less of his formidable attention. He began arriving home late in the evening, often without offering Amanda even a simple hello. His tie and walrus moustache were customarily tracked with salt and tacquito crumbs, his breath with tequila, salsa, and beer. He whispered on the telephone behind the closed door of his office late at night, and rarely slept with Amanda in their bed anymore. Instead, he took his sheets and pillows directly to the living-room sofa without issuing a single word or gesture of reproach. When she emptied his pockets at the laundromat Amanda discovered matchbook covers from unknown restaurants, breath mints, telephone numbers scratched into embossed paper napkins with faulty ball-points, and even a few stray bits of dental floss. One day Wayne left the house and never came back. He never even sent anyone round for his books.

Dr Wong asked her, 'After you vacuum the carpet, what do you do?'

'I wash all the dishes.'

'And after that?'

'I scrub the sink with abrasives. I rinse all the sponges.'

Dr Wong leaned back in his leather-bound flexible office chair, his feet propped up on his polished white oak desk. He clasped his hands behind his head.

'And what do you do when you feel sad, or lonely?'

Amanda could hear the buzzing of distant intercoms in the wide office building. 'I never feel sad anymore,' she told him. 'I never feel lonely.'

Sometimes, Amanda would hear Dr Wong start to wheeze, like the sound of mice in a wall. His mouth was open, his eyes closed behind his bifocal horn-rimmed glasses. He took another louder breath, more jagged and sudden this time.

'I want to make everything white,' Amanda told Dr Wong. When Dr Wong fell asleep, Amanda knew they could finally understand one another without this thick filthy office getting in their way. Dr Wong's desk was cluttered with papers and folders, his trash cans stuffed with Kleenex and Styrofoam fast-food containers. On the mahogany bookshelf sat a number of mugs stained with coffee, lipstick, and particles of food.

'I want the world to make sense again,' Amanda whispered. 'I want to bleach away everything in the entire world that isn't real.'

She promised them candies, cookies, and cakes, but the children never stayed very long.

'Would you like some ice cream?' Amanda asked them. She had already distributed bright paper party plates and green plastic cutlery. 'Would you like any more chocolate cake?'

'No, thank you,' the children said.

'Would you like me to read you another story? Would you like to go into the other room and watch TV?'

'No, thank you, Amanda,' the children said.

'Is there anything you *do* want? Anything at all. Don't be afraid to mention it. That's what I'm here for.'

'I just want to go home,' the children said, one after another, gazing around Amanda's bungalow at the white walls, the white chairs and carpeting, the white cabinets and bookshelves and doors. 'I think I heard my mother calling. I think I better go now.' Outside it was beginning to rain.

'If you just wait another moment,' Amanda said. She was poised on the edge of her white chair. She was trying to hear something very faint and imminent in the far white distance. 'Just wait one more moment and I promise. It won't be very long now. I'm certain it's almost here.'

THE *FLASH!* KID

Rudy McDermott's siege of the termite nest was inspired by the funny word 'attrition,' introduced to him by his birthday book, *We Were There at the Hundred Years' War*. He shoveled a moat circumscribing the infested oak log and filled it generously with Pennzoil looted from Father's outboard. The termites, busy inside their moldering apartments, exhibited no immediate concern, and Rudy dashed home for lunch. He returned a half hour later to find the insects constructing a bridge across the moat with accumulating drowned corpses, swarming headlong into the muck with a sort of conscientious frenzy. Rudy struck a match and ignited the moat. The ring of fire flashed and heat rushed his face. The fried insects smelled like burnt popcorn. Greasy black smoke lifted into the bright mountain sky, flames dwindled into the scorched earth. Rudy replenished the moat and lay back against a warm flinty hill, watching the discombobulated insects struggle and squirm in the ashy sludge. He flicked small stones at them as they carted their sizzling brethren into deep, buzzing tombs. Rudy reignited the moat and ran home for an ice cream and a brief chat with Father.

Father was out back on the raised sun deck with Mom. *Bushwah!* Father roared, and flung his newspaper over the railing. A few loose white sheets skimmed down the surface of the hill like manta rays. What's *this* I read? My tax deductible religious contributions go to providing flak jackets for Sister Maria Theresa's guerrilla forces in Uruguay! And who's Sister Maria fighting for? Subversives, that's who! And who do subversives hate most of all? Successful men like *me*, that's who!

For godsake, Mom groaned, prone on her lawn chair and bikinied, brown and glistening with oil like a very old salad. If there's one thing you sound stupid about it's politics.

Father grumbled, his face flushed. A black vein pulsed ominously in his forehead. He poured another icy margarita, sprinkled it with salt.

Termites, huh? Father said later, solaced by now with his fishing rod. He reeled in line from a spool that twitched and tumbled on the deck, and Rudy watched raptly over his dribbling ice cream. My old pal Bob Probosky and I knew all about termites. Or at least I did, yessir. When I was your age I busted open a termite nest, that's what I did. Bob was chicken, scared he'd get stung. Not me, though. I reached in and yanked out that mama termite with my bare hands, diced her for bait. She caught trout like a goddamn Gatling gun – yessir, she did! But did I let that fag Probosky have any? Nosir, I didn't! Sure I got stung. But I knew what I had to do and I did it – and *I* reaped the reward. The world's a jungle, boy. Only the toughest survive. You have to act fast if you want to make your mark on the world. You have to be tough if you want to become a successful man like your Father—

For godsake, Mom said, and reached for her sunglasses. If there's one thing you sound stupider about than politics it's *got* to be your crummy childhood.

With a sledgehammer Rudy returned and demolished the nest, pried loose sheaves of rotted wood. The mama termite was enormous, Rudy startled. Gravid and glistening, as long and thick as Father's forearm, the queen's convoluted envelope fitted snugly inside the log like the meat of some gigantic walnut. Reach in and yank it out? He would need a bucket. Rudy improvised, swung the hammer again. Pus and slime spattered his arms and face. The stench was terrible, and he wiped the sour taste from his lips. He ran away crying and crashed through bushes and a small stream. The crowd of trees stood around making shadows; birds chirped in the leaves. Rudy forced himself not to shiver, obligated by Father's nostalgic courage. He returned solemnly to the ruined nest. Termites swarmed away from the exploded queen, dragging bits of her flesh. Rudy unscrewed the lid of a jelly jar, crouched, shut his eyes. He scooped blindly at the nest and the jar made a wet *thwucking* sound. He screwed back the lid and flung the jar against the flinty hill, where it thudded soundly. Rudy's hands were sticky; he wiped them on the ground. The ground was dry and crusty and broke apart in shards. Rudy threw the flinty dirt over the ruined nest, cut more dirt loose with his bowie knife. Something metallic clanged and the knife bucked against his hand. He scraped the dirt cautiously. Metal screeched. Gradually Rudy cleared a patch of gunmetal black. The black was remarkably smooth, like the surface of an eyeball. A sense of great heaviness surfaced in his mind when he touched the buried object. Like déjà vu, abstract but firm. Patiently he uncovered the statue's entire surface. Two feet long, tubular, black and smooth and unblemished, without any markings or delineations whatever, as seamless as the skin of an egg. He struck it sharply with his knife, and the knife's point cracked. His fingers were drawn again and again across the smooth surface, as if here were condensed the enigmatic stuff of the universe. He clenched his teeth. Overhead the moon hooked vague clouds, and Rudy wondered, Who to tell? Who, indeed?

Sure, we'll take a look at it, Father agreed. Someday, someday soon. But not today, not right this minute. Right this minute there was fishing to do, imported beer to drink, Mom to bicker with inanely. That afternoon Mom drove to Tahoe and returned by dinner, her freshly dyed hair piled high atop her dry red face, accompanied by a strange noisy couple. The man was in the stock market, the woman in the Book-of-the-Month Club. The woman hugged Rudy viciously. The man said ha ha ha, what's that, young buck? A termite *how* big? I saw that movie. Jon Agar saves the world, doesn't he?

The image of the submerged, neglected statue infiltrated Rudy's dreams. They were deep black dreams without faces, a quicksand effluvium which filled his mind like molten ore, as if his identity and the identity of the statue were being inverted. The dreams encased Rudy in darkness; he felt warm, secure; his body was a vessel, hard and unimpressionable, like something fired in a kiln, like the heart of a planet, like the fine black powder he discovered inside the abandoned jelly jar the following morning. The fine kinetic powder jingled sibilantly as he swirled it around the inside of the glass, keening, eerie, celestial, like purported music of the spheres.

The first person Rudy lured to the statue took it away from him. A young surveyor had been prowling the woods for several days, unshaven, muttering, scratching himself, toting a small intricate telescope and clipboard. Rudy's approach was determinedly casual. He was learning that a child's enthusiasm is inversely proportional to the scale of adult priorities. Hey, mister. Want to see something weird? Hey, mister. It's right over here. Maybe somebody lost it. Hey, mister. Maybe there's even a reward.

Okay, okay, the young man conceded finally. Show me something weird. But then promise you'll go home, all right? Could you do that for me? Promise?

Mmmmmmmmmmmmmmm. Interesting . . . The surveyor touched the statue briefly, as if testing a hot iron. Cautiously he laid his palm flat against the frictionless surface, whistled slowly through his teeth. So heavy, he said, and clenched his jaws.

As the surveyor stared, Rudy's sanctioned enthusiasm burst free. He babbled hectically of his discovery: the doomed termites, the Pennzoil, Father's nostalgic fish bait, Mom's new hairstyle, the gravid queen, the immanent dreams and the fine black powder.

The surveyor grumbled, scratched his oily hair, scrawled something on his clipboard and proceeded to the fishing lodge.

Hey, mister – can I come? Rudy asked, was not refused.

Rudy pressed his face against the glass-paneled phone booth, breathing mist against the glass and pretending he was an enormous fish in a bowl.

Andy? the surveyor said. This is Steve. Yeah, the connection's terrible. I'm up at Caple's Lake . . . What? Dunnigan, Steve Dunnigan. No, I don't have a sister. We were in Dr Tennyson's seminar together, remember? Okay, okay – just forget it. I've found something up here you'll want to take a look at—

Here, Dunnigan said, shutting the glass booth behind him. Buy yourself some baseball cards.

Rudy accepted the quarter cordially, slipped it into his pocket, went to the lodge and bought a quarter-pound bag of beef jerky with one of the twenties from his genuine cowhide wallet. He sat on the front steps and chewed as he watched Dunnigan hurry bags and equipment from his cabin into a battered red Toyota. When Dunnigan drove off, the Toyota's flimsy clutch rattled like a marble in a soup can.

Rudy went home to dinner, rapidly consumed two steaks, a potato, no broccoli, three slices of hot cherry pie and a frozen Snicker's bar.

Upstairs in his loft he was only mildly queasy, and watched the portable television underneath his bedcovers.

He fell asleep and resumed the dreams again, awoke in a cold sweat, his stomach protuberant and growling. He slipped downstairs and managed a pair of ice cream sandwiches, returned to bed and the dreams again. It was as if his mind were being fed on a very short loop. Eggs for breakfast, four or five scrambled. Mom was pleased, offered encouragement. Another sandwich? Cookies? More milk, Rudy? Eat, *eat*! Marie and the girls are always talking about your skinny arms . . . Father said, Good for you, boy! Build those muscles – you don't want to be a skinny little wimp all your life, do you? You've got to be tough, you've got to take care of yourself in this world, boy. You think I'm not tough? Go on, then; try me. Hit me in the stomach. Go ahead, hit me. Harder. *Harder*, now! Show some muscle, boy. I've swatted gnats harder than that!

Dunnigan returned in the afternoon with a circumspect goateed man. They conferred beside the sunken statue, consulted pocket-size devices and departed in a jeep. Dunnigan returned again the following morning with more men, equipment, jeeps. Rudy visited the site daily, saw crowbars snap like Popsicle sticks, pneumatic hammers grind to a halt, strong men with ringed underarms herniate in chorus, puny forklifts roar as cables snapped everywhere. Helicopters beat over the secluded lakefront property, CB radios spluttered and squawked in the crisp mountain air. Still, the object did not budge. It would not budge. It was stubborn, heroic and invulnerable, Rudy thought. Just like Superman.

Father and Mom budged quite readily, however, packed Rudy up with the other belongings and relocated to the relative sanctity of their San Francisco mansion, where Rudy explored the daily papers with casual regularity. The initial notice appeared in the back pages of the *Chronicle*, amid advertisements for lingerie and quick-weight-loss clinics. The blurb included Rudy's name, Dunnigan's, date and location of find, difficulties encountered. A mere journalistic kernel, yet fecund, perseverant, it rooted and advanced to page two as Life Buried in Strange Object! and blossomed ultimately in front-page headlines:

LIFE BURIED IN STRANGE OBJECT!
Child Unearths Cosmic Treasure

Father and Mom began introducing Rudy to their friends as 'the little archaeologist in the family' before posting him off to bed when another reporter eventually infiltrated the party. The phone rang constantly, and Mom had the number changed. Reporters and cameramen populated the front porch, lunatics verged on the perimeters. The streets resounded with cymbals and tambourines. Bullhorns proclaimed the sovereignty of Jesuschristalmighty. *The Flying Saucer Gazette* accused Rudy of conspiring with sentient vegetable protein on Betelgeuse. Satanists dropped by evenings for coffee and, rebuked, splattered sheep's blood on the lawn, driveway and deluxe Mercedes convertible. A flurry of Dianetic brochures arrived daily with the harried postman. Red journalism complemented topical hysteria. Cosmic Statue Predicts Earthquakes!!! Jeanne Dixon Communicates with Telepathic Statue in Esperanto!!! Cosmic Boon to Acne Sufferers??? Rudy chatted happily with the interchangeable lunatics and newsmen until his family's tolerance was 'overextended,' Father's press release declared. All he can tell you, Mom shouted one day, yanking Rudy back inside, is that he found the damn thing, he gave it away and then he came right back home! Crestfallen, Rudy was denied permission to pose for the covers of *Jack and Jill Monthly* and *Isaac Asimov's Science Fiction Magazine*. For the rest of the summer Rudy was relegated to the video entertainment console in his isolated bedroom.

Dunnigan, along with the 'cosmic treasure,' was appropriated by UC Regents Berkeley. A visiting lectureship compensated the former while an elaborate wing of the Physical Research Center secluded the latter. Dunnigan appeared frequently on network news programs and the *Tonight* show, with Johnny Carson. Frankly, Johnny, we're baffled, he conceded. We can't penetrate the object's shell, but ultrasound has detected embedded proteins, minerals, rudimentary enzymes – materials implicit in the genesis of life. As I told you over dinner, the statue's shell is so dense that the molecules are virtually impacted together. Conceivably billions of years old, it's perhaps the by-product – or so contend the latest theories – of some titanic implosion, the devastating force of which would be unconscionable even in our nuclear-conscious age. At this point Dunnigan granted the unconscionable audience a winsome, ingratiating smile, like a Nobel laureate confronted by some giddy coed, and Johnny suggested they play tennis together real soon.

Rudy switched off the television. It was late. He couldn't sleep. The resumption of grammar school foreclosed upon the vanished summer like some formidable mortgage. Rudy awoke the next morning in an empty house. Dad in Rio, Mom in bed. The lunch, prepared by the maid, was folded inside a double bag on the kitchen counter. Rudy scanned the *Chronicle*'s comics page and devoured an eight-ounce box of Rice Puffies. Public concern over the statue had receded in the wake of renewed Middle East skirmishes. Rudy went to the bathroom, vomited anxiously, brushed his teeth, removed a frozen Snicker's bar from the freezer and chewed as he departed for the bus stop. Father had won the debate years ago concerning Rudy's education. He's going to public, not be a sissy. Just like me.

On the street corner Kent Crapps and Marty Femester were passing an untidy cigarette back and forth, inexpertly rolled from Bugle tobacco and parting at the seam. Rudy sat on the curb and handled his lunch bag to tatters.

Hey. If it ain't the rich kid. Hey, Crapps. Ain't that the poor little rich kid?

Sure is, Crapps said. It looks like *two* rich kids, if you ask me. Hey, fat boy. You better stop eating so much. You're liable to *explode*!

Rudy sat forlornly as he heard their approach. The wrecked cigarette bounced off his knee and he brushed at sparks.

Hey, maybe the fat boy's hungry. You think so, Crapps? You think he might like a marshmallow? There's a marshmallow, there in the gutter. It's a little muddy – but maybe the fat boy's *real* hungry.

Rudy hunkered submissively, anticipating his customary ridicule.

Hey, fat boy. Look what we fixed you to eat—

As the imperative mud-filled hand clamped Rudy's mouth, something unfamiliar activated abruptly in his mind. Something alert, canny, uncompromising.

Help help help quit it no no help help blech! Marty struggled weakly, like a small damaged sparrow. There, Rudy thought, his arm not strong so much as intent. *You* eat the mud this time. At a discreet distance Kent Crapps bounded up and down and shrieked for the police. Rudy wasn't even angry. He just wanted them to know he could take care of himself from now on. He had new responsibilities, through his discovery of the statue a sort of implied integrity. The weight of the buried statue filled the deep part of his mind. Nothing can hurt you, the deep voice confirmed, resounding in the immensity of remembered dreams that whirled, unalterable and patient, impervious and eternal.

Young men have responsibilities I don't care who started it you can't carry on like hoodlums what if everybody behaved like that I'm doing this for your own good, the principal pronounced, and down came their pants. The secretary pulled shut the office door. Rudy neither whined nor protested at his turn. He felt supremely confident, and listened to the deep dreamy monotone of the buried voice. Returning to class he met wary eyes and whispers. He ate a magnanimous lunch alone in the cafeteria and cached burps to be released later, in class, in improvisatory bleats.

Grade school was a breeze.

Ha ha ha, everybody laughed, orbiting him in the schoolyard. Occasionally Rudy grabbed the scrawniest of them – a homely, wheezing asthmatic – and twisted his limbs one at a time. He convinced the asthmatic to confess explicit sex crimes with his mother, his father, his dog. Everybody laughed and even the asthmatic grinned plaintively. You're a riot, Rudy. You are – you're the funniest guy I know. You oughta be a comedian. Rudy never once suspected himself of bullying. He was merely amusing his friends. He viewed popularity as a social obligation, like the ballot. When the bell rang, the timid orbiting boys dispersed readily to their classes and Rudy, in his own time, lumbered along behind, thirteen years old and one hundred ninety-seven pounds, and nobody told him what to do anymore. Not even his parents.

Rudy! Rudy, stop that! You *heard* me, young man! Let *go* of your mother's arm – and I mean *right this minute*, Father bellowed punily.

Damn, Rudy thought, and released Mom's red, perfumed arm. Damn if anybody sends *me* to military school, and flung the academy brochure in the trash. I'm not a failure. I will succeed. I am tough, too, and will make my mark on the world. Just watch.

Father and Mom departed for the Riviera, and left Rudy under the aegis of a flinching, reluctant maid. Just fine with me, Rudy thought. I don't need anybody. I'm happy to be me, just like they recommend on television talk shows. He deposited himself at the kitchen table and trooped through a stack of grilled cheese sandwiches as if through so many Saltines.

Rudy dropped out of school at sixteen. Father leased him a two-bedroom apartment in the financial district and promptly departed with Mom to Rio where, it was rumored, they developed a successful liaison with two blond, liquid women Mom had met in Toronto the year before. Rudy, meanwhile, ate. Mountains of toast, vistas of jelly and syrup, acres of Rice Puffies and Sugar Dongs and Candy Cakes and Twinky Pies. Crushed plastic cereal toys littered the floors of his apartment. A mobile landmark, Rudy strolled immensely through the neighborhood, easily visible from high office buildings, helicopters, incoming passenger planes. He visited Taco Heaven, Mrs Mary's Candy House, Happy Jack's Ice Cream Palace, and returned home munching candy apples, barbecued sides of beef, Big Macs. He squeezed blithely through the crowds of slim, fashionable secretaries, and never glanced twice at their slit skirts, high heels and polished nails. Desire never pestered Rudy; his pubic hair remained downy, innocent. The family doctor proposed hormonal supplementation. Adamant, Rudy refused. He was not sick. He was inconceivably healthy. His life was purposeful, coherent and determined: he ate, he slept, he waited.

Steve Dunnigan appeared at Rudy's door one summer afternoon. Rudy was uncertain of the year. The seasons had flitted by like moths. Rudy shifted his weight away from the door and Dunnigan sidled into the cluttered apartment. Dunnigan wore a faded Grateful Dead T-shirt, stained Levi's, tattered Keds. My, how you've grown, he said. Rudy slumped into a beanbag chair, and the straining plastic envelope burst with a pop, spewing brown varnished beans everywhere. Rudy sagged unconcernedly as the chair depleted, listening to the vaguely familiar man through his stuffy brain.

I came to warn you, Dunnigan said.

Rudy yawned. Dunnigan scratched his head, and white dandruff spilled on to the floor.

Have you heard about IRM, Rudy?

No, Rudy croaked, and massaged his Adam's apple circumspectly.

Innate Releaser Mechanism. Genetic knowledge, knowledge coded into the DNA. Instinct, really. But an instinct, a mechanism, which must be triggered by a behavioral cue, understand? Mother bird does a little dance, perhaps, and activates the fledgling's migratory program. Then the fledgling departs for Tehachapi, Capistrano, Guam.

Rudy reached for the crushed Ritz Cracker box, rattled crumbs into his mouth.

The cue was tactile, Rudy.

Rudy tore open the box, licked more yellow crumbs from waxed paper.

A few years ago, undergraduates at UC Research came into contact with the statue. Today these students are withdrawn, antisocial, disrespectful of authority, obese and under heavy sedation at UC Medical. The doctors and scientists have agreed on a tentative diagnosis. The prognosis is catastrophe . . . Rudy, are you listening?

Rudy picked up the telephone and dialed Chicken Delite. Three buckets of center breast, he thought, and a gallon of coleslaw. The line was busy.

The statues are containers, Rudy, distributing life's essential ingredients throughout the universe. But the molecules of the container must be fused, the container launched. Think of a simple atomic reaction. A solitary atom is split, and the devastation is well publicized. Your body is composed of how many trillions of atoms, Rudy?

Rudy put down the phone; his head lolled against the wall. A few last beans dribbled from the exhausted plastic envelope.

Cosmic evolution – just think about it, Rudy. Life is forged from calamity, catastrophe, annihilation. The ultimate purpose of life – mere perseverance. And the law of evolution? Survival of the fittest—

Father, Rudy said. Hypnagogic, he stared at the ceiling.

Rudy, wake up!

Rudy started upright. Chicken Delite? he asked.

Would you like to see the statue again, Rudy? Would you like that?

Yes, Rudy thought. Yes yes. He raised himself courageously to his feet. The varnished beans seethed on the floor.

There's food in my car. Hungry, Rudy? Come on, Rudy, come on . . . Dunnigan led Rudy out the door, rolled open the side of his van.

Rudy clambered inside, smelling pizza. Three cardboard containers streaked with oil. He opened the top box. The pizza was still warm, the cheese stiff and congealed. He divided the slices and transferred them, slice by slice, into his mouth. The van's door slammed shut; bolts were thrown. Rudy chewed pepperoni, mozzarella, briny anchovies.

The van's engine erupted, along with a nervous spasm in Rudy's stomach.

The van moved out. An air vent communicated with the driver's seat.

Everything will be fine, Rudy. They dig out a tiny chunk of your brain – no bigger than a sausage. You'll be happy, then. People will like you; you'll like people. We'll start you on an exercise regimen, a diet. Hell, with your money, you can just take your pick of the ladies. You won't be lonely anymore. You'll be just like everybody else.

But I'm *not* like everybody else, Rudy reassured himself, and placed his palm against his stomach. Something percolated deep inside; his bowels contracted. He tried to hold it in. Father would get very mad. Father hated when Rudy smelled up the car, and rolled open all the electric windows.

Just you wait and see, Rudy. We'll command top dollar from the university, once I inform them of your condition. Let me handle everything. Did I tell you they fired me from my position? I used to know Johnny Carson and his wife personally. Now what's my doctorate worth? All-night pizza deliveries to junkies, high school parties, perverts. But I've learned. This time they'll deal on *my* conditions. This time I'll demand *tenure*—

The pressure mounted in Rudy's stomach. He cried out.

What's that? Watch your temper, Rudy. I don't want you to end up like the others at UC Med. Arm straps and Thorazine – very uncomfortable. And more than anything, Rudy, I want *you* to be comfortable. The fridge at our motel is packed with Candy Cakes, Twinky Pies, Rice Puffies and plenty of that white soul food – mayonnaise and Wonder bread.

Rudy returned the final slice of pizza to the container, closed the lid. He had lost his appetite.

– Did I mention the color TV?

Rudy lay flat on his back, gripping his stomach with both hands. Just when the pain grew intolerable, the deep voice interposed. Life is light. Life is calamity, catastrophe, annihilation. You are life, Rudy. Annihilate. Annihilate color TV, Rice Puffies, UC Medical, Innate Releaser Mechanisms, the financial district, military school, the homely asthmatic, the monotone principal, marshmallows, Johnny Carson, icy margaritas, Sister Maria Theresa, Uruguay, Father and Mom. Will they see me in Rio? Rudy wondered. Just before they feel the impact of your cosmic prestige, the voice answered. Rudy chuckled contentedly. His colon fluttered.

Will they be proud? What will they think when they see me?

What the termites thought when the hammer came down. Life is light.

Every muscle in Rudy's body contracted at once. And then suddenly, just before the flash, Rudy realized he would finally make his mark on the world.

THE MONSTER

Over a period of many years, the Monster patiently assembled itself from various stray parts and pieces collected from neighborhood garage sales, thrift stores, bargain basements and swap meets. Oily gears and levers, large iron washers embossed with dark sticky substances, the ruptured transistor spines of discarded pocket radios and home intercoms. The Monster was such a perfect Monster, it often claimed to forget its own origins. Because it was large, powerful and indelicate, it called itself a man. Because it contained complex intestinal spools of memory and data, it called itself smart. As it came to perfect and complete itself, it became hard, visible, perishable and keen. It walked in the streets. It saw people and spoke with them. It noticed birds in the sky, clouds, airplanes, and the cool clinical matrices of high power and telephone lines. 'I see,' the Monster said. 'I walk. I talk. I know. I be. I am.'

In order to know the world, the Monster thought one day, I must let the world know me. This was the Monster's first philosophical thought, and so the Monster cherished it, as well as the particular sprockets and spanners from which the idea had first been gently disengaged. Shortly thereafter, the Monster went into business designing and manufacturing miniature, fist-sized mechanical Monsters for children, who bought them and their multitudinous accessories eagerly and with much fanfare. Monster House, Monster Railway, Monster Designer Jeans, Monster Sauna, Monster Barbecue Patio Set, Monster Porsche and, of course, the ever-popular Monster Family – Roger, Tina, Troy and Wendy.

'You are the sort of man who knows what he wants from life,' the Monster's secretary, Tracy Simpson, said admiringly one afternoon, just after her third piña colada. 'When you see what you want, you go out and get it. You don't beat around the bush. You don't sit around trying to make up your mind. Like, my old boyfriend, Ron, he *never* made up *his* mind. So finally I had to make up Ron's mind for him.'

The Monster and Tracy were married the following June and began raising a subsidiary family of their own – sudsidiary, that is, in relation to Man-Monster Industries, which had become so enormously successful that it was now governed by strange, distant men in tall, dark buildings with reflective windows. Cool geometries of finance widened the Monster's world, making it at once more transcendent and more real. Stock portfolios, notational diagrams, employee psychological profiles, marketing surveys, payroll and tax. 'I began this company seven years ago in order to vindicate myself to the world,' the Monster said one day, at the Annual Stockholders' Meeting in Reno. 'I now have no self left that I wish to vindicate. This is a fine company which has extended itself beyond the parameters of any one individual. It has joined itself with the world of business, numbers, profit and loss. Please keep in touch. I will keep my mail box open here at the office.'

That afternoon the Monster went home to its family, where the atmosphere had grown rather stale, gray and desultory over the years. The enormous house filled with enormous furniture. A blue pool in the yard. Topiary hedges shaped like various Monster toys and accessories. There was even a small, spurting fountain. This was a home where the Monster had never belonged. This was a family where the Monster had never been known.

We are divided by different histories, the Monster thought. The history of flesh and the history of iron. Simmering cells, protein, DNA, RNA, blood. Cotton mills, pistons, microwaves, household appliances, plumbing. Men had lived their lives for many thousands of years before iron, eating everything they could get their hands on, killing everything in sight. Sometimes the Monster wished it could know that hard visceral anger which had once helped men eat. Perhaps then the Monster could live its dense, curious life less wary of being eaten.

'Hello?' the Monster said. 'Is anybody home?'

Silence resided here, contradicting the constant, noisy beat of the

Monster's secret, clock-like interior. The Monster went into the kitchen and locked the doors. It opened and closed a number of kitchen cupboards, assembling appropriate implements on the finely varnished wormwood countertops. Then it began making meticulous adjustments on itself.

You will be more patient and attentive, the Monster said with its tools. You will take your wife dancing. You will spend more time with the kids.

'Dad?' David called from outside. The kitchen doorknob rattled. 'Dad, is that you?'

'It's me,' the Monster said. Across the countertop lay disemboweled components resembling bolts and washers, springs and flywheels, toaster grills and turkey timers. Hurriedly the Monster reached for them.

'Just a minute,' the Monster said. 'Please. Don't come in just yet.'

David didn't say anything. The Monster could feel the boy's weight in the hall, a poised and invisible attention as profound as the silence itself. The Monster's son knew the bloody world of skin and spine and lung. The Monster's wife and daughter knew. The mailman and the milkman and the gardener and the tall grocery clerk knew. Even when they didn't know they knew, they still knew anyway.

Then one day the Monster fell in love, and all its doubts and inhibitions seemed to vanish with the breeze. She was strong, caring, self-confident and, somewhere deep beneath the smooth efficient whirring of her hard mechanical parts, vulnerable and sensitive as well.

'From the moment you first arrived,' the Monster told her, 'it was as if we had always known each other. I felt I could confide in you, in your smooth planes and aerodynamic contours, in that funny, anxious little charge we share whenever we touch. And then, of course, there's the way you always listen. You're really a wonderful listener.' Her name was Amana Stor-Mor, and they shared many warm summer nights together, until that summer ended.

The Monster's lawyer and his wife's lawyer agreed on an out-of-court settlement, and the Monster lost everything. Its house, its savings account, its controlling interest in Man-Monster Industries, even its self-respect. And then, just when the Monster thought there was nothing left to lose, it lost Amana as well. One day the Monster returned home to pack its clothes and found that a strangely purring, over-equipped and falsely congenial creature had taken her place in the suddenly invidious kitchen.

David said, 'It's like, Dad, we couldn't bear to have it around, you know? Mom's detective-guy showed us the photographs and, well, it's like pretty disturbing to contemplate. It's like pretty perverted for some kid in his formative years to have to tell his friends, like, yeah, my Dad's got this thing for a frigidaire.'

The Monster carted essential possessions out to its Mercedes convertible and drove away. The smog that afternoon was palpable and rough. The Monster felt itself filling up with a sort of gray, diffuse soundlessness. At first it was a mere nanospeck, some remote decimal fragment. Soon, however, it filled the Monster's chest, face, simmering coolant, creaking levers and hidden clocks. Then, achieving the stern geometric horizons of Monsterness, the absence invested everything.

The Monster parked its Mercedes in the parking lot of a fifteen-storey downtown budget accommodation called the Hotel Cecil, the rooms of which featured oddly stained sinks and toilets, heavily painted fractured walls and ceilings, cheap plywood doors bearing the scars of many violently ruptured chain locks, and a direct view of the Greyhound bus depot across the street. Each day the Monster sat by its window peering down at its automobile in the parking lot. There is something infinite and strange in machinery, the Monster thought. Just like people, machinery often refuses to be known.

In the evenings, vandals and burglars arrived. They cut the doors of the Mercedes with bottle openers, bifurcated aluminum beer cans, crowbars, rings and knives. They disconnected hubcaps, tires, tape deck, burglar alarm, battery, hood ornament and taillights. They carted away the wheel wells and the bumpers. Soon there was little left of the Mercedes, and after the Monster's third call to the LAPD its carcass was hauled away by a grumbling and slow-witted tow truck who went by the name of Tony's Shell.

The Monster could not even suffer properly. It tried to torment itself, but only managed to invent paradoxes, word games and conundrums. It tried to debase itself, but only succeeded in figuring pi to its final decimal place. Suffering was not a lesson to be learned, the Monster thought, so much as someone to be. One suffered in order to become rooted in the very reality of oneself, like a plant in the mud.

And so the Monster began making a few internal modifications. 'Does this hurt?' the Monster asked itself, applying various tools and implements. 'Does this, or this, or does *this* hurt? How about this? Or how about *this*?' The Monster's skilled hands gripped, pried, toggled and screwed. Inside, the Monster experienced dark, tangled thoughts, dim whispering memories about people it could never be.

'Yes,' the Monster answered itself. 'That does hurt. And that hurts, too.'

Then, succinctly, it began to cry. It cried for hours and hours and, just as succinctly, stopped.

'Now I have felt pain,' the Monster thought. 'Now I have been hurt. Now I am angry.'

Because it did not feel betrayed by the world of humans, the Monster lavished its anger on the world of machines.

'No pain is as terrible as my pain,' the Monster told the world. 'No anger is as righteous as my anger.' Then it crushed them. It crushed Monster Man and Monster Woman seated at their miniature dinette set on the cracked particle-board bureau. It crushed Monster Children at play on the four-piece Monster Playground and Recreation Area. It crushed the entire Monster Family as they were setting out to the beach in their Monster Limo, replete with Monster Picnic Basket, Monster Pet and Monster Beer Chest. Soon the Monster's minimal apartment was filling up with more and more Monster Rubble. 'Arr,' the Monster cried, perambulating clumsily towards the Monster Family Yacht as it floated in the sink's tranquil, greasy water. 'I hate machines, I hate toys, I hate Monster Families!' Then it lifted the helpless boat out of the water. 'Help, help!' it imagined the Monster Family shouting. 'Please, help us, please!' The Monster smashed each Monster doll against the edge of the sink; their heads cracked open like eggshells, their various limbs snapped sharply like twigs and toothpicks. 'I hate Monsters!' the Monster roared. 'All Monsters must die!'

By this time, one of the Monster's neighbors usually began beating the walls or ceiling with a bedboard or umbrella.

'Keep it quiet in there!' the neighbor shouted. 'What are you – *crazy* or something?'

One warm spring afternoon someone knocked at the Monster's door. The Monster did not respond right away, for many of its response-mechanisms were rusty or frozen over with disuse. It waited to hear another sequence of knocks. Then, slowly, methodically, the Monster arose from its creaking bed. As it crossed the room, the fragmented plastic frames and steel machinery of broken Monster toys crunched under its feet. The Monster opened its door.

'Like, hey, Dad. Like, it's me. David. Your son, right? Can I come in, or am I supposed to stand out in this hall all day? I mean, you don't look half bad. Your place is a dump, but you yourself don't look half bad at all.'

David was twenty-five already, and vice president in charge of his own manufacturing firm in Hermosa Beach. David was very blond, very tall, and very well groomed. 'I thought we'd have a meal together, you know? Have ourselves a little Father-Son-get-reacquainted sort of reunion and all that? You could tell me how you've been, what you've been doing, and how you could just walk away like that from your own flesh and blood, your own children, like we didn't even matter – oops. Sorry, Dad. Let's have that meal. I promised myself I wouldn't start in on you like that.'

Like suffering, the Monster learned reconciliation – a softer, more subliminal sort of pain, one requiring less-severe modifications. The Monster ordered a pork *machacha*, David a chicken burrito. David recounted his two failed marriages, and previous addictions to cocaine and Valium. His third wife was older, with two children of her own. Meanwhile, the Monster ordered more salty margaritas.

'Like, Dad. Do you mind? I've always wanted to ask you like a question?'

'What's that, son.'

'Like, what are those wires coming out of your neck?'

Self-consciously, the Monster touched them.

'My neck,' the Monster replied, sensing a vague dissonance. David was already regarding him with an unformed expression.

'Oh,' David said after a while. 'Like, I always thought you were just wearing one of those Sony Walkmans or something.'

And so the Monster went to live with its son and daughter-in-law, its step-granddaughter and step-grandson, where the house was always filled with the human beat of heart and blood, like the kick and recoil of tiny levers and tiny springs, as primordial and unremitting as the pulse of any clock. This was the conspiracy of cells and molecules the Monster had always hated itself for not knowing before. Children opening and closing refrigerator doors and kitchen cabinets. Men, women and children from other houses arriving and departing. There was something ominous about the children and the way they always watched. They liked to touch the Monster's face and hands. They palpated its skin and musculature, and measured its nostrils with plastic rulers and compasses. They asked it ceaseless questions and were never satisfied with any of its answers.

'How does a television work, Grandpa?' they asked. 'Or a radio, or a car? How does electricity keep some things cold and other things hot? Where do microwaves come from, and what is the greenhouse effect, anyway? Does that mean the polar ice-caps are melting? Or does it just mean the way plants manufacture oxygen?'

They were always asking, asking. Crawling into the Monster's lap when it sat reading the paper, running their soft hands across its harder, more efficient ones. It was as if they suspected. It was as if they knew precisely what functions in the Monster they could compel. Sometimes they just looked the Monster silently in the eye, and these times bothered the Monster most of all. They wanted to know what it was thinking. They wanted it to answer the question that was itself.

The Monster watched television with them in the artificially warm rooms. It corrected their math and science homework. It posed in awkward positions on the shag-carpeted floor and permitted them to climb over and around its superficial structure. It read them stories from large cardboard picture books filled with urgent primary colors. Pictures of polar bears in stocking caps and mittens, bespectacled gophers and hibernating geese conferring around tattered maps and spinning compasses. It even read them illustrated adventures of Monster Man and Monster Family, since Man-Monster Industries had recently expanded into almost every area of consumer merchandising. There were now Monster Man bath products, and Monster Woman feminine-hygiene deodorants.

'Monster Man and Monster Woman married and were very happy together,' the Monster read out loud, pointing at the pictures of the man, then the woman, in the book. 'They moved into Monster House, and parked their Monster Car in the big, echoing Monster Garage. Then one day Monster Woman went to Monster Doctor, and came home and gave Monster Man the good news.'

The Monster paused for effect, looking from Child Number One to Child Number Two. Both of them were rapt and breathless, filled with the pulse of their own sinewy, parenthetical hearts.

'Do you know what that surprise was?' the Monster asked, glad to be the one who was asking the questions for once. 'Do you know what Monster Woman learned at the doctor?'

Child Number One looked up. His attention seemed at once penetrating and opaque, as if he sought to render everything, even himself, transparent. 'Were they going to have a baby?' Child Number One asked. Child Number One was the boy, because he wore blue pajamas and blue socks.

'That's an absolutely correct response,' the Monster said. 'They were going to have a little Monster Baby.'

Then, sometime in late winter, the deep structure of Child Number Two began to malfunction. All day and all night Child Number Two lay staring in her pink bed, among her pink, illustrated blankets and stuffed toys. There was always a damp, milky residue in her eyes and in the corners of her mouth. She never smiled anymore, and took all her meals in bed. She didn't laugh, or run, or play. Child Number Two was beginning to function very ineffectively in her role as the family unit's youngest non-financially productive child-member. The Monster could always tell Child Number Two was the girl, because she wore pink T-shirts and pink socks.

Doctors and therapists began making housecalls. They plied Child Number Two with medicines, ointments, conditioners and diet supplements. Three days a week Child Number Two's mother drove her to the hospital for special treatments that left rough, scabrous rashes on her chest and throat; eventually, her hair began to fall out. The Monster's son, David, began pouring his drinks from large recurring jugs of Albertson's Vodka, and David's wife, Mary Lou, cried abruptly from time to time, without any reason at all, even in the middle of her favorite programs on TV. Whenever the Monster tried to question them, David and Mary Lou flashed with sudden anger and reproach. Mary Lou carried her clutched handkerchiefs into the bedroom, slamming doors and windows along her way.

'Don't act like it matters to you, Dad,' David said, taking more long drinks from his tall glass. 'There's nothing you can do, so why bother – isn't that *your* attitude? You never bothered about anything that ever happened to me in your entire life.' Then David disappeared into other rooms, too. Whenever David disappeared, the gallon jugs of Albertson's vodka always disappeared with him.

One night, simply out of curiosity, the Monster entered Child Number Two's room. While she slept, it performed its own cursory examinations. Child Number Two was malfunctioning at the level of cells and enzymes. Tiny electrons misfired. Bodily fluids were redirected by trauma, cellular shock and unscheduled hormonal rushes. The Monster thought of pistons, solenoids, the circulation of oils and lubricants. This stuff called blood, it was a problem. This heart and this brain. The Monster went downstairs to the basement for its tools. Then it returned to Child Number Two's bedroom, where everything was glowing and indistinct.

The Monster pulled back the blankets and removed Child Number Two's pink pajama top. Integument, blood vessels, striated muscle and calcified bone. The Monster closed some major arteries with paper clips; it triggered a number of pain-suppressant levers with alcohol-swabbed hypodermic needles. This was what the Monster most disliked about human beings. Blood was so much messier than metal.

The Monster installed tidy machinery in Child Number Two's heart and spleen. It scraped out a nervy gray malignancy with an X-Acto knife. Then it cleaned everything with alcohol and solvent, and closed Child Number Two's chest with a needle and thread from Mary Lou's sewing basket. Suddenly, there was something comfortable and embracing about this human house. The world of blood is faulty, frail and adjustable, the Monster thought. Just like the world of metal.

Before the Monster finished closing her up, Child Number Two's eyes opened abruptly, like the weighted glassy eyes on one of her plastic dolls.

'Grandpa?' Child Number Two said.

'What.'

'I think I'm starting to feel a little better.'

'I know,' the Monster said. It leaned over and bit the thread, tying the loose ends together. 'I know you're better. Now let's get these sheets down to the laundry before your mother wakes up and sees this mess.'

Night penetrated the basement washroom with a fine ambient darkness. The Monster stuffed bloody sheets and blankets into a large, very obvious Maytag, including appropriate measures of detergent and bleach.

'Grandpa?'

'What?'

'Tell me a story. Tell me a *good* story.' Child Number Two was sitting in the sofa-chair beside the boxes of bundled newspapers, unraveling tennis and racquetball racquets, neglected photo albums and Family League bowling trophies.

'What sort of story?' the Monster asked. It engaged the wash cycle and rested its hand on the thrumming metal lid. Appropriate lights activated on the otherwise expressionless panel of knobs and indices. The basement is always here, the Maytag said. Me and the basement. We'll always be here.

The Monster turned to Child Number Two, who was touching the tight stitches across her thin chest as if they were the strings of a cello. 'What sort of story do you consider a *good* story?' the Monster asked.

'The story about the Monster who went to Mars,' Child Number Two said quickly. She was squirming already, as if blocks of remembered narrative were a sort of fuel. 'Or the Monster who saved the world, or the Monster who loved the princess when the princess didn't love him. The Monster who won the war, or the one about the ugly Monster who grew up to become a beautiful, tall office building. Tell me the one about the Monster's underwater city, where he makes friends with the gigantic octopus and the menacing black squid. I want to hear stories about when the Monster does good things, and makes other people happy. I want to hear stories where the Monster makes lots of friends, and provides for his family, and lives happily ever after in the Monster City, hidden away in Australian jungles where he can't be discovered by the world of men. I want to hear one of those stories, Grandpa. I want to hear one of those stories before I go back to bed.'

The Monster heard the Maytag shudder and shift into rinse. The darkness was beginning to disassemble, invaded by the cool morning light. The Monster heard creaks and soundings in the upstairs rooms and hallways. Feet on the floor. A discreet toilet flushing.

The Monster sat down and placed Child Number Two in its lap. Child Number Two was a strange, unappreciable sort of creature, even when she was functioning correctly. She was like quantum mechanics, or the creatures that drifted alone in the depths of unfathomable black oceans.

'Okay,' the Monster said after a moment, checking Child Number Two's pulse with the flat palm of its hand. 'But remember – these are not the only stories.'

GREETINGS FROM EARTH

Dear Diary,

Well here I have gone and done it, and have decided to keep a diary at this point in my lifestyle because I think the many 'out of body' experiences I have been experiencing lately will be of much interest to many different people. Which is not to mention my second reason for keeping this diary, which is that my husband Roger Simpson never pays any attention to my feelings not even one iota, of which this (my diary) is a direct result I think. Sometimes I get so mad at Roger I can't see straight I'm so mad sometimes. Right now Mr Genius is hammering in the basement on our family room, only he's been hammering on this supposed family room for two years now and I don't see much of a family room. All I see is one big disaster area filled with wood and rusty nails and Bosco, Roger's stupid dog he never walks, has peed and crapped on practically everything. I don't know what most people imagine when they imagine a supposed family room, but big yellow puddles of dog pee is not the first thing that comes to my mind, that's for sure.

'Out of body' experiences by the way are highly interesting moments which relate to ourselves all the time, of which you are not even aware sometimes since they sneak up on you for instance. At first you feel a sleepy sensation in your toes. Then you see a soft white light and feel very peaceful all of a sudden, as if you did not have a care in the world. Then you hear this faraway music which sounds sort of like Mantovani only better. Then you feel lighter than air and before you know it you are experiencing your 'out of body' experience at that very moment! Maybe you are lying in bed reading a good article in the *TV Guide* or maybe sitting on the living-room sofa. Maybe you are washing the dishes or maybe vacuuming the rug. Only now you are a disembodied essence standing apart from your physical vessel and looking at it, almost like looking at yourself in a mirror, only now you are free to wander the world in your highly spiritual state and you feel better about yourself than you ever have had previously before in history.

Being in a disembodied essence situation has many highly interesting advantages one can often enjoy. First of all you are on a much higher intellectual plane than ever before, i.e. you're a lot smarter. You're even smarter than Roger 'Einstein' Simpson, your stupid husband. Your stupid husband and your stupid house drive you crazy now, since housework and marital duties all seem terribly demeaning to your spiritual essence, and you would prefer to go out to a good movie or even dancing. Maybe you put on some nice clothes and go for a walk, or maybe you go across the street to the Kona Lanes Bowling Alley which has a nice bar and pretty good oar derves for Happy Hour, not that you care much for these material things since being in a disembodied essence situation means you are never hungry and never thirsty and you never have to go to the bathroom.

Roger of course would never believe you are a disembodied essence in a million years probably, since he has told you a million times already he does not think you are a very metaphysical person, which of course is now a big joke on stupid Roger. Roger thinks a person cannot be highly spiritual unless they read dozens of crazy books all the time like Roger does and try to understand all the metaphysical problems which bother his mind every day of the week. Sometimes these problems keep Roger awake all night and so he goes downstairs to work on our family room, where he says eventually we will have a color teevee and maybe even a pool table. The main philosophical problems facing mankind today Roger says is the mind-body problem and the decay of the earth's ozone layer problem, which two problems Roger says are totally connected to each other all the time. Roger tried for many years to solve these problems and get a Ph.D. at UC Irvine, but unfortunately his professors told him he was a 'nut-basket' and would not help him one iota in any way, shape or form. Sometimes I tell Roger maybe he should try to get some sleep and stop worrying so much about problems which don't make him very happy or put any bread on the table, but then Roger says I am just like all his old professors and that I cannot understand anything which is not grossly materialistic. Roger says I have what Roger calls a 'conventional moral outlook' which makes it impossible for me to see the major issues facing our planet today. People like Galileo, Christopher Columbus and Melvyn Dumar will always have to fight back against people with 'conventional moral outlooks' Roger says if our society is to make any decent progress anywhere.

Sometimes I don't think Roger is very metaphysical at all but maybe that he is just emotionally disturbed or something. Or maybe he has one of those seventeen-foot worms eating his brain like I read in the National Star News one time. The man in the National Star News story was a big shot atomic scientist, which just goes to show that even having some brainiac IQ doesn't always protect a person from having a worm eat up his brain. But honestly, even if Roger was completely disturbed or something I don't know anyone who would believe me since everybody, my family included, seems to think Roger is a real genius and all and that I have made quite a 'catch' in getting him 'to the altar' so to speak.

Anyway, I have written enough for one day and so I will stop.

Here for example are the sorts of books Roger reads every day of the week.

Immanuel Kant, The Metaphysics of Morals and The Critique of Pure Reason.

Hegel, The Phenomenology of Spirit.

Wayne K. Shabullah, Know Your Egoic Ray. (Roger by the way says Wayne Shabullah has likewise received much trouble from colleges with 'conventional moral outlooks.' Roger says Shabullah is the only other person in America besides Roger who has noticed that egoic rays are often misdirected by satellite interference, which means many people are often accidentally aiming their egoic power at their own self-image factor or libido-release mechanisms, which thereby weakens people rather than their opponents in the social battle struggle for proper genetic improvement of our species. Or something like that. Personaly though what I think Roger and Wayne K. Shabullah have in common is that neither of them has both oars in the water, if you will pardon my expression.)

John Locke, An Enquiry Concerning Human Understanding.

Wayne K. Dwyer, Pulling Your Own Strings. (This one I have read a little of, as well as seeing Dr Dwyer on The Merv Griffin Show, whose book actually makes a lot of good hard common sense unlike Roger's other books.)

I'm afraid though none of these books has done Roger's brain any good at all since his ideas just keep getting weirder and weirder. Today for example Roger tried to tell me that his mind-body problem is a lot like the family room he's building in the basement. Roger says the idea of the family room he has in his brain is perfect, but making this idea perfect in reality is another story entirely. I try to tell Roger that nobody in the world really cares if our family room is perfect or not, particularly since we don't have any children or any friends who still visit our crazy household. Frankly Roger I tell him I think you're just putting a lot of pressure on yourself which you don't need right now, particularly at this point in your life. But of course Roger isn't listening to a word I say, since he is staring at my television program instead which is The Merv Griffin Show, with Merv's special guests today Don Rickles and Gloria Loring. Roger says television is just the sort of problem he is talking about, since nobody on television is real at all but instead is just invented by the worldwide global corporation conspiracy in order to enslave the spirits of mankind. The worldwide global conspiracy wants people to watch television all day long so they cannot see the world around them is being secretly bought up by a group of Taiwanese businessmen, whose next step is to transform the world into a giant petrochemical storehouse which will provide life-giving vitamins and sustenance to invading life-forms from the sixteenth dimension. These invading life-forms (and the Taiwanese businessmen) are seeing to it that Earth is filled with pollutants in order to transform the way men's minds work. Pollutants and petro-chemicals create mental imbalances in our cerebral mechanisms, Roger says, because the mind is not a free place but like the family room depends upon its own material existence to survive in the world all the time. That is why once Roger has made the family room just like the one in his mind he will prove to himself and all free-thinking individuals everywhere that mankind can stand against the worldwide global conspiracy no matter how much they fight us, which of course by the time Roger has finished telling me this I have missed practically all of Merv Griffin.

Then of course Roger goes downstairs and starts hammering again, and I could probably have more peace and quiet in an insane asylum probably. Personaly I think Merv Griffin is maybe not a big genius like Roger, but still I think he is a very nice man with a nice sense of humor, and I like him because you can tell he is not the sort of man who goes home and gives his poor wife a lot of grief about how she may not be the proper intelligence quotient for him and all. I like to watch Merv because after work I am very tired and need to unwind, and at least he is not another noisy cop show or another western show. Thank God especially he's not another Bonanza. Bonanza's on just about every station on cable about ten times a day. I hate Bonanza.

I am getting to the opinion that my life would make an interesting movie, being that my adventures in the world of spirit could teach many interesting things that should be shown to the world. I have decided to write my experiences into a screenplay, and have decided the best person to play my part in the screenplay would be Jessica Lange, who's movie Frances I have seen now twice already and understand it very well, boy do I. In it (the Frances movie) Jessica Lange portrays a very beautiful and intelligent woman who gets put into a mental asylum by her mother even though she's not the one that's crazy but actually it's her mother that's crazy.

The first scene should take place in the home where Jessica Lange lives with her crazy husband Roger. The house is filled with books and papers and dirty dishes everywhere, because Jessica has been at work all day making money to put food on the table, and Roger is sitting on the sofa wearing his crummy green and yellow Balboa High School jacket which he never lets Jessica wash, not that washing clothes is the big thrill for Jessica Roger seems to think it is. Jessica Lange's mother is sitting with Roger in the kitchen eating the rest of the pot-roast Jessica was going to warm up for dinner. Jessica Lange's mother should be played by the same woman who played her mother in Frances, being they are both the highly critical sorts of persons who never say good things to a person but only highly critical bad things. When Jessica opens the front door she is exhausted from working all day and carrying two bags of groceries three blocks from the store, because of course she and Roger can't afford a car as she is the only person in their household with a job and all.

ROGER: Being a highly intellectual person myself I think everybody in the world should worry about things like the mind-body problem and the world global corporation conspiracy, because as very intelligent people like Kant and Hegel say all the time, blah blah blah blah blah blah blah.

JESSICA LANGE'S MOTHER: Roger, this is very interesting and I sure enjoy listening to your brilliant mind in action, especially since you are probably the smartest man in the world, probably even a genius or something. Jessica Lange doesn't have any idea how lucky she is being married to a man with such a high intelligence quotient, since she herself has never been very intelligent at all, and didn't even finish high school due to pregnancy—

At this point I think Jessica really blows her top. She has had a lousy day at work and then had to do the grocery shopping and buy a new Lady Bug lady's shaver at Sears, and now the kitchen is filled with dirty dishes and big bags of garbage because her husband Roger has never picked up a dirty dish or a garbage bag in his entire life let alone wash one. Normally Jessica is a very spiritual sort of person but that doesn't mean she has to take any nonsense, not even from Roger. She throws her groceries on the ground and says

Just shut up Mom you stupid idiot. Just because I don't sit home all day reading books doesn't mean I'm stupid. Personaly I would much rather have some passion and excitement in my life, and I don't just mean sex (not that I'm doing so well around here in *that* department) but I mean passion and excitement in a very spiritual sense too. I would like to be able to afford a decent vacation every once in a blue moon though, or even a cute little silver sports car – maybe even a convertible, and if that makes me too materialistic for an Einstein such as yourself, Roger, you can just go and fly a kite. There's nothing wrong with having a few nice things that make you happy and it doesn't mean you can't be spiritual at the same time.

As you can see, Jessica really lets them have it. She is too strong to let anybody treat her like some stupid cow. Every time her husband Roger starts talking about his crazy ideas, she will scream at him,

'Earth to Roger! Earth to Roger! Come in, Roger!' and the entire audience will laugh because Roger will look like one really stupid jerk. Even though she is a very spiritual person who has many 'out of body' experiences, Jessica knows how to behave reasonably to people and keep her feet on the ground when she has to.

Being a disembodied essence is important for many different reasons, some of them being that you don't need anything or anybody to make you happy, not even money and not even a man. A lot of the time I watch my physical vessel going through its daily routine and I can't believe how stupid I use to be. Every day my physical vessel gets up at the crack of dawn and takes the bus to Taylor Morgan in Anaheim where it types up invoices for public lavatory appliances. Then it goes to the store and comes home. Then it cleans and cleans, without its husband even coming upstairs for one second to give it a kiss or a hug. Then it eats Cheese Puffs and watches television. Then maybe it takes a Valium and eats some Nugget-sized Reese's Peanut Butter Cups, which come in a large assortment bag and must have about a jillion calories apiece in them. Then maybe finally it lies down and after a while it falls asleep. I feel very sad sitting at the vanity table and watching it toss and turn without anybody to hold it or tell it it's one bit special. I know you are not the happiest person in the world, Helen, I want to tell her. You were raised without any proper father to speak of, except for your mother's stupid boyfriends, most of which were stupid jerks. You are not what they call attractive by any means, and you are also heavy around the middle. You can't even finish the *TV Guide* crossword puzzle which you were working on before you fell asleep, and everybody knows that the *TV Guide* crossword puzzle is the easiest crossword puzzle in the entire world. You had a baby when you were in high school and had to give it up for adoption because your stupid mother made you, which means the only positive thing you ever did in your entire lifetime (i.e. giving birth to another human being) was just a waste. You have always given your life away to other people, and now you hardly have any life left at all. Even when you become peaceful and spiritual it is only in your mind, and so your poor physical vessel doesn't even know what being a disembodied spirit feels like, or how many advantages it has over normal material existence all the time.

My physical vessel doesn't even hear a word I'm thinking about, even though it has woken up and reached for the Reese's Peanut Butter Cups again. This makes me feel very sad and depressed, and so after a while I dress up in my blue dress and even wear my nice high heels. I go out for a walk to the Kona Lanes bar where Armando the bartender tells me I look pretty, which even though I know Armando is just lying to me to be polite makes me feel good anyway, since some men know a lady likes to hear a nice compliment about her appearance every once in a blue moon and not just a lot of mumbo-jumbo about the seventeenth dimension and all.

Sometimes I go away for a few days. Sometimes I even get a room at the Kon Tiki Motel and put it all on Roger's Visa card. I spend the whole night watching television without having to listen to Roger's crazy hammer all night through. Sometimes I order room service, and sometimes I go to a first run movie. I wonder why it has taken so many years of my life to find happiness, and if there are any problems in life I will ever solve. Sometimes when I think about how difficult life is or how sad my physical vessel must be at home in its lonely bed I feel very sad, and sometimes I even feel a little sorry for Roger. Maybe somewhere in Roger's nutty brain he feels sad sometimes too, and all the big problems he wants to solve are somehow related to finding a way not to be sad anymore.

Whenever I come home again there is my physical vessel getting along perfectly well without me, going from one dull routine job to another just like a robot. While my physical vessel sprays the greasy gray spots on the refrigerator with Formula 409 I sit down at the kitchen table and try to figure why I bother coming back to all of this at all, and I am stumped. On the kitchen table there is a box of Clairol hair coloring which my physical vessel bought for itself while I was away. My physical vessel is wringing the sponge in the sink and humming 'God Rest Ye Merry Gentlemen' even though this isn't even close to Christmas, but is actually only September.

Finally I go downstairs where Roger's hammering. The entire basement is one holy mess, with dog hairs and pee and crushed Cherry Coke cans everywhere. Bosco starts whining and wiggling his butt like he thinks I'm going to hit him for peeing everywhere, but I understand why he does it since Roger never lets him out in the backyard even for a minute. I tell Bosco to go away though, because Roger never brushes him and whenever I pet Bosco I get dog hairs all over my clothes and face, not that my clothes are so special or anything.

Roger is hammering thousands of nails into all of the wooden wall supports. When I come in he says, 'I'm experimenting with nail-frequency. Nail-frequency means the spacing and design of nails in order to discover the most efficient panel attachment effectiveness ratio. The Parthenon was built both to last *and* look good, you know.'

The nails are shiny and in many different patterns, and by this time I am thinking Roger should probably be put in a big boat and pushed into the Pacific Ocean. The nails are shaped in double lines, curlicues, waves, circles and semicircles. I want to tell Roger that his family room looks a lot like Glen Campbell, but then I decide not to bother since Roger never gets any of my jokes at all, not even my good ones. Roger picks up a copy of 1001 Home Ideas from the workbench and shows me the cover. 'I've decided once I've finished our family room I'll build one of these.' He points with the hammer. 'A window box for the bathroom.'

I roll up the bag of Gravy Train which is on the stairs and put it on the radiator, and then I sit down on the stairs. 'First of all Roger I want you to listen to me very carefully because I won't be telling you any of this again. I am not really your wife Helen at all, but instead I am her disembodied essence which has decided to leave its physical vessel forever in order to wander the world on its personal quest. I don't know what my personal quest is exactly, but it's probably to find out who I am since nobody, not even my mother, ever seemed very happy with me, which certainly never gave me a very good self-image factor.'

Roger puts down his hammer, which must be for about the first time in a million years probably. Then he opens another can of Cherry Coke. 'Spirit is not such a great place to be all the time,' he says. 'Spirit has always gotten all the good press though, so naturally people are often confused. Actually the spirit can't be happy unless it has a nice home in the material world. The material world, and I think most people will agree with me on this, is a very nice place to be most of the time.' He shows me his can. 'The material world has Cherry Coke.'

'I don't like Cherry Coke,' I tell him. 'I like Tab, because I am trying to cut down on my calories.'

'The material world has shopping plazas, Chrysler LeBarons,

Hostess Ho Hos, Mitsubishi, Saran Wrap, Johnny Walker Black Label and Sears.'

'I can still go to Sears,' I say. 'Being a disembodied essence doesn't mean you still can't go to Sears.'

'You can't buy anything.'

'Yes you can.'

'You can't drink Tab. You can't eat Quarter Pounders. You can't you know what.'

I can't believe how stupid Roger is sometimes. 'Mere physical gratification is not the most important thing in the world, Roger. And as far as you know what, I couldn't do much worse in that department, and that's for sure.'

'I should take you out more. I should take you to a movie.'

'You can take my physical vessel to a movie, Roger. I'm getting out of here.'

'I should take you to dinner. Maybe even dancing. I really hate dancing, but maybe I should.'

'You can take my physical vessel dancing, Roger. It's upstairs now watching *WKRP in Cincinnati* and eating Cheese Puffs. I'm sure it would appreciate a night out on the town.'

Roger reaches for his hammer. Then he places a nail against the wall and squints at it, as if he is aiming a rifle. 'We'll do a movie this week. And next week we'll talk about dancing.'

I get up from the stairs and brush the dog hairs off my legs. 'Goodbye forever, Roger. I don't bear you any bad feelings, and I hope you find peace with your own inner-self like I have done. I hope you solve the mind-body problem, and that you exercise and eat better and stop drinking so much soda. And someday I hope you sweep up all this dog hair. This basement is a mess.'

'Okay, honey.' Roger isn't even looking at me. He seems to be thinking about something important. Then he starts hitting the nails again with his stupid hammer.

That's about all I can take. I walk upstairs where my physical vessel is wearing the blue flannel housecoat it bought for itself last week at K Mart. I don't look back, and leave that crazy house forever.

Life is a very funny thing I guess, which is an expression you hear often.

Roger and my physical vessel seem to be getting along better than ever without me since Roger has decided to go back to college again, even though he is almost forty, because he has gotten a teaching fellowship so he can do his part to educate the youth of America about today's big social problems, in which Roger is an expert of course.

Now that Roger is back in school he has forgotten all about his crazy family room, and every day after work my physical vessel sits and watches Merv Griffin without anybody to bother it while Roger is at the library finding more books to bring home. Yesterday I stopped by to see how my physical vessel was doing, and it sat on the sofa staring at the television and eating Cheese Puffs for almost two hours without moving, just getting older and fatter.

But now back to my spiritual quest, of which my screenplay is drawing to a close already. The ending will be very positive and upbeat so the people watching the movie will feel good and not feel like they wasted their hard-earned money. Jessica Lange's spiritual essence gets a good paying job as a computer programmer and rents a nice apartment in Beverly Hills after selling her first ever screenplay to the movies. Her boyfriend is named Dirk Stevens who she has met at a big Hollywood bash, and he is a real lady's man who goes out with all sorts of different women, but once he starts going out with Jessica she 'lays down the law' and he decides to give up all his other lady friends for her, and after they are married he tells his friends he never thought he'd settle down in his entire life, but that was of course before he met Jessica who is a very special sort of woman who won't take any nonsense from men. Jessica takes night classes at UCLA and eventually gets her Ph.D. in creative writing, at which all her professors say she is a natural, and one professor is so impressed that he has an affair with Jessica and falls in love with her, but Jessica doesn't think much about it because Dirk is the only man she can ever truly love. Her ex-husband Roger, though, lives alone in his big house with his books, and he's never really happy because he realizes no woman in the world will ever treat him as good as Jessica used to, and he drinks and drinks and drinks every day until eventually he becomes an alcoholic and dies.

Well that's about the end of my personal tale I guess, and now that I am free of the boring nightmare of my physical life I thank my lucky stars all the time. Every day I feel like jumping for joy, or going out to dinner for a really good steak, because spiritual love of myself has taught me how to enjoy all of life and not just some tiny part I happen to be stuck in. Sometimes late at night for example I even get on the RTD bus and travel for hours and hours without even having to pay any extra money, and many of the drivers know me by sight already I am sure. We go to Santa Monica, Encino and even Orange County, where I have never been before. I am very beautiful these days, because true beauty is knowing yourself as a spiritual being and not letting anybody fill you with a lot of negative feelings about yourself, which is about all my stupid life with Roger ever did. Now that I am filled with love and beauty I feel very content and warm in my little seat in back of the bus. Sometimes I feel so peaceful it is like I am not even there at all, and there is only the city outside with all the streetlights and office lights on, and my face shining very clearly in the window. I finally feel in my heart of hearts I am free of that nutroll Roger for good, as well as that crazy house, Bosco's dog hairs and working and cleaning seven days a week just to get nowhere fast. Anytime I feel like it I can go dancing. I am free to travel to many foreign countries and make many new and unusual friends. I can spend weekends at Club Med which I read about in the Sunday Travel Section. Inside I am just like Jessica, dancing with Armando out on Harbor Boulevard like a crazy person and letting everybody in the world watch, and what do I care. I take rafts down the Colorado River, visit darkest Africa and help feed starving babies in poor countries. I travel through space like a beautiful angel and walk on the icy moon. Outside the streets are dark and empty, and the bus is roaring underneath me, and it feels like there's nobody else alive in the entire world.

Sometimes I can travel secretly like this for hours and hours without anybody ever seeing.

UNMISTAKABLY THE FINEST

Every time Sandra Mitchelson's daddy came home on the boat he brought her things. French chocolates, a stuffed elephant, a golden heart-shaped locket, a transistor radio, a hand-painted porcelain Japanese doll with rice paper parasol. In return Sandra helped him work in the backyard. The front yard was covered with gravel, the backyard with tall yellow weeds. 'This will be our family area,' Daddy said, knee-deep in the weeds. 'We'll have a barbecue, a swing set, a birdbath, a trellis, maybe even someday a swimming pool.' They already had a fishpond. The water was dark and smoky, rimmed with algae. Large gold- and lead-colored fish glimmered dully in the muck, slowly blinking their bulbous eyes like monsters surfacing from some nightmare. Sandra held Daddy's white cloth hat and watched him hit the ground with a shovel. He overturned convexes of damp black earth, severed worms and pulsing white slugs. Sandra liked the pungent, musty odor of the fertilizer, and rode on Daddy's back while he pushed the reseeder. They watered every morning, and soon tiny green shoots appeared. After Daddy disengaged the garden hose he filled his coarse, red hands with water from the tap, flung the water into the bright summer sky and told Sandra the sparkling droplets were diamonds. Sandra tried to catch them, but they slipped through her fingers. One day she sat down on the patio and cried. Daddy promptly took her to the store and bought her a tiny 'Genuine' brand diamond set in a thin copper band. The next morning he went away on the boat.

The new grass died, the earth turned gray and broken. Mrs Mitchelson said, 'He wants a lawn? Then let him water it his own self, why doesn't he.' She toasted her reflection in the twilit picture window. 'Here's to your damn lawn. Here's to your family area.' Bourbon and crushed ice spilled over the rim of her glass. In the afternoons Sandra sat alone on the living-room floor and observed through the smudged picture window the gradual destruction of the yard. In the spring, weeds grew – strange enormous weeds as tall as Daddy, bristling with thorns and burrs and furred, twisted leaves. Scorched by the summer sun, the weeds cracked and fell, and when spring returned, the mat of dead weeds prevented new weeds from sprouting. Sandra asked when Daddy would be home. Mrs Mitchelson said, 'Never, if I have anything to do about it,' and departed for the pawnshop with the heart-shaped locket, the transistor radio, the tiny 'Genuine' diamond ring. 'You want to know what all that junk was worth?' Mrs Mitchelson shouted, looming over Sandra's bed at three a.m. Sandra sat up, blinked at the light, rubbed her eyes. Mrs Mitchelson's eyes were red and wet and mottled with discount cosmetics. 'Twenty bucks. That's how much he loves you. Your wonderful father. Your father who is so wonderful.' Mrs Mitchelson stormed out of the room, the front door slammed. Sandra rolled over and went back to sleep. That summer they sold the house.

In Bakersfield Mrs Mitchelson worked at the Jolly Roger Fun and Games Lounge next door to the public library. Every day after school Sandra waited in the library and read magazines. She especially liked the large, slick magazines that contained numerous full-page advertisements. She enjoyed reading phrases such as 'unmistakably the finest,' 'the affordability of excellence,' 'the passionate abandon of crushed velour.' When the library closed at nine she sat outside on the bus bench and thought about the sharp, clear photographs. Fashions by Christian Dior, natural wood-grain furniture, Chinese porcelain, a castle in Spain, a microwave oven with digital timer, an automobile with a leopard crouched and snarling on the hood. The doors of the Jolly Roger swung open and closed, releasing intermittent bursts of smoke, laughter and jukebox music. Buses roared past. Sometimes one of Mrs Mitchelson's friends drove them home. Nervous, unshaven men; their cars were usually littered with plumbing or automotive tools; cigarettes with long gray ashes dangled from their mouths. They ate pretzels and laughed with Mrs Mitchelson in the living room while Sandra went quietly to bed.

*

They lived in Pasadena, Glendale, Hawthorne, Encino. Sandra finished high school in Burbank, acquired a receptionist's job in Beverly Hills. In Compton they took a one-bedroom apartment which included some cracked windowpanes and numerous discreet cockroaches. Weekdays, however, Sandra sat at an immaculate mahogany desk in the public relations firm of Zeitlin and Morgan. She answered telephone calls (often from television and film celebrities), organized the week's appointments in a large leatherbound black ledger, typed advertising copy and allowed clients into the security building by activating a hidden white buzzer.

Sandra was usually alone in the office. Mr Zeitlin had retired to compose Bermuda postcards. Mr Morgan – with his distinguished gray hair, taut polished cheekbones and jogging outfit – arrived each day around elevenish, then quickly departed with Elaine, the leggy secretary, for the afternoon luncheon appointment. Occasionally Mr Morgan's son Matthew dropped by and asked for Elaine. 'Off with the old man again, huh? When am I supposed to get *my* chance?' Sandra admired Matthew – his capped white teeth, his knit ties, his shirts by Pierre Cardin. He resembled a man of 'casual elegance,' sipping Chivas Regal on a sailboat, displaying Jordache emblems at garden parties. Matthew was an executive with the Jiffy-Quick Messenger Corporation of Southern California. His solid-gold tie clasp depicted the comical (but fleet-footed) Jiffy Man dashing unflappably to his appointed destination. 'My Dad didn't just hand me the job, either,' he assured Sandra. 'I started off at the bottom, and absolutely refused any sort of preferential treatment. I even drove the delivery van one weekend, so nobody can say I didn't pay my dues. It literally took me months to get where I am today, and it was never any picnic, let me tell you. But I like to think that in the long run my employees will respect me for it.'

At four o'clock Sandra pulled the plastic jacket over the IBM, replaced paper clips and memoranda in their appropriate drawers and locked the office. On her way to the bus stop she window-shopped along Rodeo Drive, observed silk crêpe de Chine slacks at Mille Chemises, solid-gold Piaget quartz crystal watches at Van Cleef and Arpels. She admired the white, unblemished features and long, cool necks of the mannequins; their postures were perfect, their expressions distant and unperturbed, as if they attended a fashionable cocktail party at the heart of some iceberg. Maseratis and Mercedeses were parked along the curbs, and elderly women in low-cut blouses walked poodles on stainless-steel leashes. Everything and everybody appeared immaculate and eternal, like Pompeiian artifacts preserved in lava. Sandra avoided her own reflection in the sunny window fronts – her pale white skin and shiny polyester skirt made her feel like a trespasser in a museum. She caught the six-fifteen bus and generally arived home just after dark.

Mrs Mitchelson started awake at the sound of Sandra's key in the lock, sat bolt upright on the living-room couch. 'Who's that? What do you want?'

Sandra opened the hall closet, removed a hanger. 'It's only me. Go back to sleep.'

Mrs Mitchelson's dry tongue worked soundlessly in her mouth; she cleared her throat. 'Well,' she said experimentally. 'Well, I wish I *could* go back to sleep. I wish I *could* get a minute's peace around this place.' She gripped the frayed arm of the couch with both hands and pushed herself to her feet. 'But don't worry about me. Just because I gave birth to you. Just because I took care of you when *you* were sick and helpless.' Mrs Mitchelson took three short steps and landed in the faded rattan chair. The chair creaked sympathetically. 'I'm not saying I was perfect. I'm not saying I didn't make my share of mistakes. But at least I *tried* to give you a good home – which is sure a hell of a lot more than your father ever did.'

'Sit down, Mom. I'll get your dinner.'

'Do you think it's easy for me? Do you? Getting older and weaker every day, so sick I can hardly breathe sometimes. Just sitting around this lousy apartment wondering how much longer I've got left in this miserable life.'

'Please, Mom. Don't say things like that.' Sandra folded the comforter and slipped it under the couch. 'Do you want Tater Tots or french fries with your dinner?'

Mrs Mitchelson's attention was diverted by the TV tray which stood beside her chair. The tray held a depleted gallon jug of Safeway brand bourbon, an uncapped liter bottle of Coca-Cola and an unwashed Bullwinkle glass. 'Why not? Why shouldn't I say it? I hope I *do* die. I hope I die tomorrow – how do you like that?' Mrs Mitchelson absently cleaned the glass with the sleeve of her blue flannel bathrobe. 'You wouldn't miss me. You'd finally be free of me, just like your father.' She filled Bullwinkle waist-high with bourbon, adding a few stale drops of Coke for texture. 'When I needed your father, where was he? Traipsing all over the world, *that's* where he was. *You* might as well be a thousand miles away too, for all the good you ever do *me* . . . *Ah*.' Mrs Mitchelson put down the empty glass and snapped her dentures with satisfaction.

In the kitchenette Sandra turned on the stove and emptied a can of Spaghetti-O's into a saucepan. She could hear the neck of the Safeway jug clink against the rim of the glass.

'When I remember when I was younger, all the opportunities I had. I had a lot of boyfriends. They took me to nice restaurants, bought me expensive presents. Then I met your father. I was so stupid stupid stupid. I threw everything away for that louse. *Now* look at me.'

Bullwinkle looked at her.

The following summer Mrs Mitchelson was admitted to City Hospital. 'This is just what you've been waiting for, isn't it? Now I'll be out of your hair for good.' Mrs Mitchelson's voice was uncharacteristically restrained. Sometimes she almost whispered, leaning toward the side of the bed where Sandra sat. 'But just you wait. Now you'll learn what it's like to be alone. You'll know the hell I went through when your father left me for some cheap Filipino whore.' Mrs Mitchelson's eyes were wide and clear and moist, like the eyes of Bullwinkle on the drinking glass. Sandra sat quietly with her mother behind the cracked plastic partitions, listened faintly to the moans and cries of neighboring patients, read paperback romances in which elegant women were kidnapped and fiercely seduced by pirates, rebel cavalry officers, terrifically endowed plantation slaves.

Mrs Mitchelson's cirrhosis was complicated by undiagnosed leukemia, and she died unexpectedly just before dawn on a Monday morning. Sandra was fixing coffee in the kitchenette when the nurse called. Her mother had been wrong, she abruptly discovered. She did not feel alone, she did not feel betrayed. She did not, in fact, feel much of anything. She took the morning off from work, arranged disposition with the hospital crematorium and smoked a pack of Mrs Mitchelson's menthol cigarettes.

The medical bills were formidable, and Sandra had less money than ever at the end of each month. Her window-shopping expeditions grew less frequent, and she began taking an earlier bus home. Without Mrs Mitchelson to care for she rarely thought to fix dinner. She became pale and listless. Elaine said, 'Why don't you lunch at Ramone's today? They've got an outdoor patio and it's a beautiful day.' Instead, Sandra remained in the office alone, lunched on vended crackers, bagels and candy bars.

Then one night Sandra discovered Reverend Fanny Bright and the Worldwide Church of Prosperity. Reverend Fanny's sermons were broadcast live every Saturday evening from Macon, Georgia. Reverend Fanny told her followers, 'You can't expect happiness to just come *knocking*. You must *pursue* riches, you must *pursue* happiness, you must *pursue* the power of Divine Creation. When you see something pretty you want to buy, how many times have you told yourself, "I cannot afford this"? Is *that* what you think, children? Is *that* what you believe? Then you are *negating* the power of Divine Creation. You must convince yourself you can afford *anything*. You *can* afford it, you *will* purchase it, you *shall* possess it. You must impress your super-conscious with *affirmation*. The super-conscious is His workshop where, with the divine scissors of His power, He is constantly cutting out the events of your life. But first you must show Him the *patterns* of your desire, you must fill your *mind* with beautiful things.' After each sermon Reverend Fanny pulled a chair up close to the audience and solicited tales of miraculous prosperity. Middle-aged men and women described flourishing investments, sudden cash gifts from strangers on the street, gratuitous office promotions. 'All I want to tell you,' one woman said, 'is that I love you, Reverend Fanny. Prosperity has taught me how to love. Now I no longer feel so empty and alone.'

Every month Sandra mailed the church a check for ten dollars. In reply she received a mimeographed request for further donations. The stationery was inscribed with the church motto: *If you do not wish to be denied riches, you must not deny riches to others*. Sandra closed out her savings account, transferred the $2,386 to her previously minimal checking account and prepared herself for imminent prosperity. She purchased navy cashmere sweaters, suede pants, a silk crêpe blouson dress fringed with lace, a deep-breasted brown satin coat, labels by Calvin Klein, Oscar de la Renta, Halston, Adolfo, Bill Blass, Ralph Lauren. She joined a health spa, subscribed to tanning treatments, visited prestigious beauty salons. Her checking balance dropped to $1,900, $1,350, $1,000. She did not question the beneficence of Divine Creation; instead she used her Visa card. Elaine said, 'You're looking so much better, girl. Why don't we have a drink together after work? I'm meeting a couple of Tokyo software executives, they told me to bring a friend.' Mr Morgan granted Sandra a fifty-dollar raise and told her, 'You really bring a lot of class to this office,' on his way to a toothpaste manufacturer and lobster bisque.

Church doctrine was unequivocally validated. Sandra increased her monthly contribution to thirty dollars.

Then one night Sandra discovered her super-conscious in a dream. She ascended a long winding staircase. She was wearing her ankle-length 'Cameo lace' nightgown from Vassarette, her Nazareno Gabrielli padded cashmere slippers. Her fingers ran lightly along a polished oak banister. The summit of stairs met a long off-white corridor lit by globed ceiling fixtures. The fixtures were white, opaque and sprinkled with the silhouettes of mummified insects. At the end of the corridor a solitary door stood slightly ajar. Bright yellow light from behind the door cast long, angular shadows down the length of the corridor. Sandra stepped quietly, afraid of disturbing anyone. As she approached the door she grew light-headed and her ears popped, as if she were descending in an airplane. The tarnished aluminum doorknob rattled at her touch. She pushed open the door.

The room was small, windowless, lit by a naked overhead bulb. Cobwebs scribbled the pale wall and cornices. The plaster was pitted and crumbling, mapped by an extensive network of cracks and crevices. The hardwood floors were sagging, whorled and discolored. A full-feature model A–20 integrated amplifier sat in the middle of the floor beside a matching AM-FM stereo digital-frequency synthesized tuner and cassette player. An identical system had been advertised in *Stereo Review*, and Sandra still recalled many of its vital statistics. A pair of three-way loudspeakers were stacked against the wall, with twelve-inch woofers, four-inch midrange drivers and one-inch dome tweeters housed in walnut-veneer cabinetry. A mass of electric cords were joined by a plastic adaptor to a solitary wall outlet. A tiny green light activated on the amplifier's monochrome panel, an eight-track tape clacked faintly inside the tape player. The speakers suffused the room with white, cottony static.

Louis Armstrong began to sing, accompanied by bass, piano and drums.

Baby, take me down to Duke's Place,
Wildest box in town is Duke's Place,
Love that piano sound at Duke's Place...

Sandra disliked jazz, pulled shut the door. The music diminished to a low, persistent bass that fluttered in the off-white corridor like a staggered pulse. The door's surface was Formica, with simulated wood grain. She tested the knob, the lock clicked distinctly. Then she woke up.

It was still dark when the music awoke Sandra on the living-room couch. She reached sleepily for the portable television. The green, baleful screen stared vacantly back at her, containing only her dim reflection – a shrunken body attached to one enormous, elongated hand. Louis Armstrong continued to sing.

Take your tootsies into Duke's Place,
Life is in the swim at Duke's Place . . .

The bass thudded soundly in the floors, the walls, the cracked wooden frame of the couch.

Sandra turned on the lamp and saw the stereo components stacked against the far wall, partially hidden behind the TV tray. The amplifier's monochrome panel glittered intricately. I am a miracle magnet, Sandra thought, recalling one of Reverend Fanny's prescribed affirmations. Beautiful things are drawn irresistibly to me. I give thanks that every day and in every way I grow richer and richer.

On her way to work Sandra mailed the church a check for one hundred dollars.

That night she couldn't sleep. She lay on her back on the couch, her hands folded on her stomach. She closed her eyes and tried to visualize the off-white corridor, the half-open door. What did she want to find inside? A color TV, jewelry, kitchen appliances, a new car? What kind of car, what color? Would it fit inside the room? How, exactly, had the room looked? She remembered the pitted walls, the stained floors, the quality of light – but she couldn't put all the elements together at once. A Maserati, she decided firmly. Like the one Mr Morgan drives. There, it's all decided. Now she was closing the door. Okay, the door is closed. Everything is very dark. Had she heard the living-room floor creak just now? Yes, she was almost certain. Still, she kept her eyes closed a few more minutes.

She sat up and opened her eyes. The living room contained the portable TV, the aluminum TV tray, the new stereo, the broken wall clock, the dingy venetian blinds.

She closed her eyes and tried again. No matter how hard she concentrated she could not make the car appear. It was nearly dawn before she fell asleep and stood again on the winding staircase. The wooden stairs were firm and cold against her feet; they even creaked occasionally. The car, she wondered. Will the car be there, or something else? It doesn't matter, she told herself. She would accept what was given. She wasn't choosy; she wasn't greedy. She only wanted her fair share. She walked to the end of the corridor, pushed open the door. Books were stacked haphazardly around the small, otherwise empty room. Dozens and dozens of books, as if waiting to be shelved by some divine librarian. Sandra stood at the doorway, but she did not go inside. The room's strange powers might harm her, she thought – jolt her like electricity, singe her like fire. She pulled the door shut.

When she awoke the next morning, she examined her new books. They were accompanied by a bright orange and green brochure which described them as 'The Greatest Books Ever Written.' *Madame Bovary*, *The Scarlet Letter*, *Fathers and Sons*, *The Red and the Black*, *Jude the Obscure*. Each volume was bound in genuine leather and filled with numerous illustrations by 'The World's Greatest Modern Artists.' She imagined the spines upright and glistening on a brand-new bookshelf. A blond oak bookshelf, perhaps. With glass-paneled doors and gleaming gold fixtures . . . But any bookshelf will be fine, she reminded herself abruptly. Really, any kind at all. She wasn't in any kind of hurry. She didn't want to test the power, or challenge it unduly. She would accept what she was given.

Every night the dream recurred and the room presented her with beautiful things. A Schumacher 'Pride of Kashmir' Indian rug, hand-carved Japanese console with iridescent moiré lacquer wash, a hand-cut glass chandelier by Waterford, a Miró original, a Roe Kasian dining set, a Giancarlo Ripa white-shadow fox fur. The next time Matthew visited the office she was wearing Fernando Sanchez's latest, a sheer silk-taffeta dress anchored to a black lace bra. Her ruby earrings were the color of pigeon's blood. Matthew sat on the edge of her desk.

'You like Japanese food?'

Sandra stopped typing, looked up. Her lashes were Borghese, her mascara Lancôme. 'I guess I don't know. I've never had it before.'

'Never?' Matthew's face was puzzled, as if confronted by an enigma. 'Tempura, teriyaki, Miso soup? You're in for a real treat. I know the best place in town. They've got shrimp the size of my fist.' He showed her his fist for emphasis. 'How does eight sound?'

'Eight?'

'All right. Eight-thirty – but try and be on time. I'll only honk twice. Here.' He handed her the steno pad. 'I'll need your address. Draw me a little map or something.'

Matthew picked her up at nine and they drove directly to his apartment, a West Hollywood duplex. 'Is the restaurant nearby?' Sandra asked. For the occasion she wore an obi – a broad black sash belt – with her cobalt-blue, raw-silk dress. 'It just suddenly occurred to me,' Matthew said. 'They probably aren't open Thursdays. I'm almost certain, in fact. If you're hungry, see what's in the fridge.' In bed Matthew was fastidious. His hands and mouth made routine, scheduled stops at each of her erogenous zones, like miniature trains on a track. Sandra, meanwhile, observed herself in the mirrored ceiling. 'What's the matter with you?' he asked finally. 'You didn't tell me you had problems with men.' Matthew's body was sleek, firm, unblemished. His underwear was by Calvin Klein, his cologne by Ralph Lauren. Sandra said she just wanted him to hold her, and Matthew grew suddenly tense in her arms. He said he was short of cash at the moment – could she pay her own cab fare home? He *would* reimburse her.

Matthew stopped coming by the office. Whenever Sandra called his home she couldn't get past the girl at his answering service.

'Matthew Morgan residence – Mr Morgan is out at the moment. Can I take a message?'

At this point Sandra usually heard the click of a second extension being lifted, and knew Matthew was listening when she asked, 'Has he picked up his messages today?'

'One second and I'll check . . . This is Sandra again, right?'

'Yes.'

'Well, I'm afraid he still hasn't called in. But you can leave another message, if you like . . .'

One day Sandra waited outside Matthew's office building until he emerged for lunch. 'You know I really care about you,' he said. 'I just think it would be better if we didn't see each other again for a while. It's nothing the matter with you, babe. It's *me.* I just don't think I'm ready to make the kind of commitments you seem to expect from a man. You tend to be very possessive – which is *fine*, it's only *right* . . .' He paused to wave at his secretary, who tapped one foot impatiently at the curb. 'Look, baby. Let's talk about this later in the week, okay? We'll have lunch. And do you think I could borrow a twenty until then?' He palpated his vest pocket. 'Seems I left my wallet in the office.'

Matthew never called. Sandra waited at home, certain he would. She broke dates with Mr Takata, the software executive, and Steven, her aerobics instructor. It was only a matter of time. Matthew would come around. She was a miracle magnet. She was one with the creative power. One night she received an Amana trash compactor, the next a Zenith Gemini 2000 color television. 'You must not be afraid of total fulfillment,' Reverend Fanny warned, her brows knit with sincerity. The glazed, speckless TV screen crackled with static electricity as Sandra reached to increase the volume. 'You mustn't fear, you mustn't doubt, you mustn't lose faith. *Total* fulfillment requires *total* commitment. Have you, for instance, hoarded away a little nest egg, some rainy day money? Then you doubt the complete power of Divine Creation. Why put a time lock on your security savings when your love can be bullish on the stock exchange of heavenly devotion?'

That afternoon Sandra walked to the corner and mailed the church a check for $327.43, the balance of her account. Later the same night, Matthew called.

'Hello, Sandra?'

'Yes.'

'Sandra, baby. It's me. Matthew. You remember me, don't you?'

'Of course I do.'

'I know I should've called. But it's been really hectic in the messenger biz, you know?'

'I'm sure it has.'

'You're not mad or anything, are you? I seem to sense a lot of hostility on your part. I know I owe you twenty and all—'

'No. I'm not mad. I'm glad you called. I was waiting for you to call.'

'Good. Look, I was thinking. Let's have dinner tonight – all right with you? I'll pick you up a little after eight, we'll go find a nice quiet spot.'

Matthew arrived at half past ten, and rang the doorbell.

'You think my car's safe parked outside? It's so late, I thought we could scrounge up a snack right here. You can show me round your apartment.' Matthew entered the living room. 'Hey, where'd all this great stuff come from?' He reached for the tape player – *Louis Armstrong's Greatest Hits*. Lengths of crumpled brown magnetic tape spilled on to the floor. 'Seems you've got the tape caught on the heads. If you've got a screwdriver, I can probably fix it.'

'I pushed a wrong button or something,' Sandra explained quickly, took the tape from his hands and plugged it back into the player. 'I'll have it fixed one of these days. I just like the way it looks. It really brightens up the room, don't you think?'

'What's down here? Is this the bedroom?'

Sandra followed Matthew through the door. He crouched in the corner of the bedroom, picked up and assessed one silver candelabrum. 'This is worth a few bucks,' he said.

'It's getting late. Aren't you tired?' Sandra asked, and began straightening the Wamsutta silk sheets.

'Where'd you get all this loot? My old man doesn't pay you this well just to answer telephones.'

'My father sends me things. My father has a very important job in Asia. Now, please – put those things down. Get into bed.'

'What a sweet deal. I think I'd like to meet this old man of yours someday.'

Sandra pushed Matthew's hand away from her belt. She just wanted him to hold her, she said. This time, he obliged.

Sandra and Matthew were very happy together for a while. She enjoyed cooking his meals in the microwave, washing his clothes in the Maytag. Every morning she walked to Winchell's and brought him coffee and jelly donuts. Matthew took the next few weeks off from work. 'I want to be with you more,' he said, wiped a dollop of red jelly from his chin and peered over Sandra's shoulder at the new Panasonic Omnivision VHS video recorder with wireless remote. 'I never saw that before. Did it just arrive this morning or something?'

Every evening Sandra stopped by the market on her way home. Matthew requested steak, swordfish, veal, king crab, champagne, Jack Daniel's Tennessee sour mash. She began computing her checking balance in negative numbers. When she arrived home Matthew was usually on the phone in the bedroom. ' – yeah, Bernie. It's me, Matt . . . I know I haven't been home – I don't see what it matters to you where I'm staying. I want you to put a grand on Blue Tone in the sixth . . . Bernie, don't insult me. You know my old man's good for it.'

Sandra collected soiled glasses and plates from the living room. On the burnished mahogany coffee table she noticed a tiny soft white mound of powder centered upon a small rectangular mirror. A gold-plated razor blade, attached to a fine-link silver chain, lay beside it. She took the dishes into the kitchen, started hot water in the sink, wiped a bit of fried egg off the lid of the trash compactor. Her mail was stacked on the countertop. A lengthy, itemized Visa statement. The landlord's second eviction notice. Urgent utility bills with bold red borders. My mind is centered in infinite wealth, Sandra reminded herself, and opened the last envelope. *Dear Friend*, the letter began. *Are you prepared to receive the wealth of Divine Creation? Then you must be prepared to dispense wealth to others. Wealth flows two ways, not one, thus maintaining universal harmony*. The letter was concluded by Reverend Fanny's mimeographed scrawl. Sandra removed her check-book from the kitchen drawer, computed her balance on the Texas Instruments Scientific Calculator. Zeitlin and Morgan would pay her Wednesday. Perhaps she could deposit the paycheck in time for her outstanding checks to clear. Her balance, then, would be $23.97. She thought for a moment, turned off the sink faucet, dried her hands on a towel. I am a money magnet, she thought. Every dollar I spend comes back to me multiplied. I have all the time, energy and money I require to accomplish all of my desires.

She wrote the church a check for one thousand dollars. She licked and sealed the envelope, then heard the broiler door squeak open behind her. She turned.

'Porterhouse, huh? Great, baby. My favorite.' Matthew slammed the broiler door shut. 'Listen, I need to ask you a little favor – *por favor*? Just a couple hundred for a day or two. My accountant's got all my assets tied up in some sort of bonds or something. I don't really understand all the technical details. It'll take me a few days to get hold of some free cash. You know these accountants. They think it's their money, right?'

Matthew's smile was beautiful. His teeth actually sparkled, like the teeth in television commercials. Matthew and Sandra are very, very happy together, Sandra thought. Marriage, they both realize, is inevitable. They mean so much to each other. They will honeymoon in Brussels, where Matthew has important family. After a year they will return to the States where, with the aid of a personable nurse, Sandra will raise two beautiful adopted children, a girl and a boy. Matthew will eventually be recruited into politics. 'We need you,' his influential friends will say. 'You're the only man who can beat Patterson.' Matthew will win by a narrow margin, but his reelection four years later will come in a landslide. They will rent a Manhattan penthouse, and Matthew will commute to Washington.

'I can write you a check,' Sandra said.

One morning before she left for work Sandra made a long distance call to Macon, Georgia. 'Worldwide Church of Prosperity,' the receptionist said. 'How can we help each other?'

Sandra asked to speak with Reverend Fanny, and the receptionist said, 'Oh, I'm afraid that simply isn't possible. Regretfully, the Reverend's numerous personal and public commitments make it virtually impossible for her to speak privately with each and every one of her brethren. But should you, perhaps, be contemplating sizable donations – say, ten thousand or more, and all of it tax-deductible, of course – then I *might* be able to connect you with one of the Reverend's close advisers—'

'I *am* comtemplating sizable donations,' Sandra assured her. 'I *am* grateful to the power of Divine Creation. My life is abundant with beautiful things. But at the moment I'm experiencing some problems of *cash flow* . . .'

'Oh,' the receptionist said.

'I'm sure it's just a temporary problem – but I was wondering if there weren't any special prayers or affirmations for someone in my situation. You know, prayers which might *focus* my miracles a little more. And please, don't think I'm trying to be greedy or anything—'

'Cash is not wealth,' the receptionist said. 'Money only travels in one direction. True wealth flows both ways. Now, if you would like to give me your address, I'll see to it you receive our free monthly newsletter.'

On Friday Sandra received a series of overdraft charges from her bank, and a tense telephone call from the local Safeway manager. 'I realize these things happen,' the manager said. 'Have to admit even I've bounced a few in my day. But how soon can I expect your check to clear?' Very soon, Sandra answered. She would deposit funds first thing tomorrow morning. She was so embarrassed. It wouldn't happen again.

'You'll get your damned money!' Mrs Mitchelson used to shout, after the store's third call or so. 'What's the matter with you people, anyway? Don't you realize I'm just a single woman, trying to raise a child? No – *you* listen for a minute. You men always expect us to listen to you – well, *you* listen for a change. I'll pay you when I'm good and ready, and not a minute sooner. Got *that*? And *another* thing. You've got the worst stinking produce section in the city, do you know that? Your apples are wormy, your lettuce is wilted, your vegetables are rotten. Do you hear me? *Rotten*. Instead of me paying you, you should pay *me* for all the lousy produce I bought from your store and then had to throw away. That's what *I* think.' After Mrs Mitchelson hung up the phone she would fix herself a drink and tell Sandra to go pack her suitcase. 'We're going to stay with your Aunt Lois again for a while,' Mrs Mitchelson would say. 'Then maybe we can find a new home where the bastards will let us live in peace.'

After Sandra hung up the phone, as a sort of grudging memorial to her mother, she climbed a stool, reached the Safeway jug down from a high dusty shelf and poured herself a drink. She carried her glass into the living room, which was crowded with mismatched and extravagant furniture, video and stereo components, unopened crates of records, Abrams art books and glassware, like the award display on some television game show. Matthew was playing Galactic Midway, an arcade pinball machine by Bally. Bells chimed, lights flashed, hidden levers pumped the next gleaming silver ball into position. Gripping the machine's sides Matthew nudged it from time to time. 'Have a good day at work?' He pushed the reset button. On the scoreboard the digits clacked noisily around to zero.

'It was okay.' Sandra cleared some stray pearls from the ottoman and sat down.

'Did I tell you the electric company called? Something about a last notice. I think I'd look into it if I were you . . .'

On the coffee table the mound of soft white powder had nearly doubled in size, like a miniature avalanche. Sandra sipped her drink and glanced around the room.

After a while she asked, 'Where's my new VHS? My video recorder. The one I got this week.'

The flippers clacked noisily. Then Matthew hit the machine with his fist. 'Damn!'

'The VHS. I asked what happened to it.'

'How the hell do I know?' Matthew pressed the reset button. 'Do I look like the maid or something? It's around someplace. You probably haven't looked hard enough.'

'I don't see it. It was in the living room this morning. It couldn't just get up and walk away.'

'It'll turn up, you'll see. Everything turns up eventually,' Matthew said, and pumped another ball into the game.

That night Sandra sat down and wrote a letter on her IBM Selectric.

> Dear Reverend Fanny,
> Please excuse the fact my check didn't clear. I had a very bad week last week. I tried to call and explain but the lady who answered the phone said you were very busy. I will try to make the check good at some date in the near future. I agree that if I expect to receive riches I must not deny riches to others, but I'm afraid my boyfriend Matthew whom I live with sold my Tiffany silverware set yesterday while I was at work in order to pay his gambling debts. Also he says he took my tape deck in to be repaired but I doubt that seriously. Also my electricity is being shut off tomorrow unless I pay them which I can't, and then what good will my new TV or any of my new kitchen appliances be good for? I would appreciate any help or advice on these matters you might like to impart to me. I wish I could send you money like I usually do but I'll try to send you twice as much next time and hope you understand and forgive my present fiscal situation.
>
> Yours faithfully,
> Sandra Mitchelson

When Sandra fell asleep later that night she did not dream of the corridor. She dreamed instead of vast darkness, where silence filled everything like a heavy fluid. The fluid filled her mouth, throat, lungs. Breathing was impossible. I believe, she thought. I believe, I believe, I believe. She looked up and thought she detected, at the surface, a glimmer of white light. She tried to push herself up through the black weight but something gripped her ankle, something warm, pulsing, insistent. Like a tentacle, it moved up her leg. 'Baby,' the darkness said.

Sandra started upright in bed. The bedroom glowed with dim moonlight.

'Baby,' Matthew said again. His arms wrapped themselves around her waist, his hands pulled her back into the weighted darkness.

On Monday, when Sandra returned home from work, she found Matthew in the bedroom, packing his Cricketeer wardrobe into the Samsonite. 'I think I've done my part. I can honestly say I've done my share to try and make this relationship work.' Underneath his packed clothes the tip of a silver candelabrum glinted dully. 'But I'm the type of guy who demands a certain amount of honesty from a woman. Once she starts lying to me I know it's time to hit the road.'

'I never lied,' Sandra said. 'I sent the electric company a check, just like I said. It must have gotten lost in the mail—'

'I'm not talking about that and you know it. I'm not talking about the fact that there aren't any lights or any food in the house – or even that the television doesn't work. I'm not talking about what it's like living in the goddamn Stone Age, I'm talking about simple honesty – something you obviously know nothing about. I'm talking about the check you gave me for Bernie, which wasn't worth the paper it was printed on. I'm talking about my reputation in this town, which is now just about shot because of you.'

'I'll make it good,' Sandra said. 'It won't be any problem. We can sell the television, the washing machine—'

'It's just a little too late for that, Sandra. Bernie went to my old man for the dough. I'm free and clear. I've still got a job to go back to, or don't you remember. You didn't expect me to live *here* all my life, did you? In *Compton*?' Matthew latched the suitcase and swung it off the bed.

'You have to stay. You can't leave,' Sandra said over and over again as she followed him down the hall and watched him walk out the front door.

The telephone was disconnected, the water, eventually the gas. Every night the vast, liquid darkness displaced Sandra's dream of the miraculous corridor. Elaine said, 'You look like hell, girl. When was the last time you took a bath? There's a distinct odor creeping into this office, and don't think Mr Morgan hasn't noticed.' At the end of the week Sandra returned home and found the front door sealed shut by the Sheriff's Department. The lock had been changed. She jimmied open the bedroom window, the one with the faulty latch.

Everything was gone. The brass bed, the pinball machine, the Maytag, the Lenox crystal. Rectangles of dust marked the former locations of impounded furniture. The room grew dark, and Sandra went to the window, turned open the venetian blinds. Outside, it was dusk. She watched the phototropic streetlamps glow and gradually brighten, casting pale, watery red light through the blinds. Now she had nothing, and it didn't surprise her one bit. She was stupid, she never did anything right. Mrs Mitchelson was right, Reverend Fanny was right, Matthew was right. Everybody was right, everybody except her. She was all alone; she was afraid of total commitment; she was dishonest – dishonest with herself. She was sick of being wrong all the time. Things must change; things were going to be different now. *I* am going to be different, she thought. From now on *I'm* going to be right, *I'm* going to make the right decisions.

She just needed one more chance. Finally she knew what it was she wanted, and that was the important thing. It was all very simple, really, like psychoanalysis on television. She wanted someone who cared about her, someone who would stay with her. Staring out the window at the streetlamp, Sandra leaned against the wall. Eventually she grew sleepy and closed her eyes.

She heard the door open behind her, the crack of its ruptured plastic seal.

Sandra opened her eyes. 'Matthew?' She turned around. The door stood open, the living room remained empty. She walked to the door and looked out. The air was filled with blinding, devotional light. I am one with the creative power, Sandra reminded herself. I am not afraid. I believe, and I am not afraid. She stepped outside.

She stood again on the winding staircase. As she ascended she turned and caught a brief glimpse of the downstairs room. The light swirled and dust motes revolved slowly, like nebulas and constellations in some twilit planetarium. Large wooden packing crates were stacked everywhere, their lids nailed shut.

Sandra reached the summit of stairs. At the end of the corridor the door stood slightly ajar.

Quickly she crossed the length of the corridor, flung open the door and, without a second thought, stepped inside.

The overhead bulb flickered and extinguished with a sudden pop. She was not afraid, she told herself. The place was very cold and very dark. Slowly her eyes adjusted. Tall yellow weeds surrounded her, rippling as an icy breeze blew past. The foundation of the fishpond was broken and upthrust; all the water had drained away, leaving a few green puddles of algae. The skeletons of the monstrous goldfish, partially devoured by stray cats, lay strewn about the yard like weird leaves. She got down on her knees. Burrs and thorns scratched her legs. Her hands groped among the weeds, discovered fragments of the Japanese porcelain doll. The tattered rice-paper parasol was damp and stained with mildew. She heard a noise and looked up.

A tall figure stood between her and the half-open door.

'Daddy?' Sandra asked.

Another sudden breeze blew past. The door slammed shut.

'Isn't that just what I should've expected.' The dark figure approached, briefly stumbled. '*Damn* – I could break my leg on these lousy gopher holes. Just look what a holy mess your father made of this place. But who's the first person you hope to see? Your father, your wonderful father who never called, who never wrote, who never came to visit, who certainly never provided one nickel of support. Your wonderful father who never really gave a good goddamn whether either of us lived or died.'

Sandra sat down on the damp ground, weeds brushing against her face. The porcelain fragments crumbled apart in her hands,

'Aren't you a little old to be playing in the dirt? Here, get up.' Mrs Mitchelson offered her hand. Sandra took it, pulled herself to her feet. 'Try and grow up a little, will you? I can't keep my eyes on you every minute. Just *look* at this mess.' Mrs Mitchelson slapped the dirt from Sandra's knees.

They took one another's hand. Mrs Mitchelson's hand was cold and dry and soft. Sandra squeezed it tightly against her stomach, afraid of the dark.

'Try and remember that sometimes I need a little help and consideration too, you know. I can't do everything. I can only do the best I can, that's all. The best I can. Come on, now, and fix my dinner. It's been ages since I've had a decent meal in this dump.'

Then, together in the deepening darkness, they made their way carefully across the ruined yard toward the shadows of the house.

DIARY OF A FORGOTTEN TRANSCENDENTALIST

March 13

Oh it was a beautiful day in the woods today. A really, really beautiful day. I really mean it. It was probably one of the nicest days I remember in a long time. Really nice and sunny. Boy, I wish every day could be half as nice. That would make life practically perfect.

March 17

Another really nice day. I took a lot of long walks in the woods and had a lot of very interesting ideas about life, nature, the universe, you know. Some of the best ideas I've ever had, in fact. But then like an idiot I forgot to write any of them down, and by the time I'd gotten back to my cabin I'd forgotten all of them.

(Note to myself: From now on carry a notebook!!!)

March 18

Remembered to carry a notebook today, but I didn't really come up with any good ideas.

March 21

A Nature Poem
by Henry Willoughby

~~Oh vast and radiant grandeur of bough and willow and pine~~
~~Oh heady clouds of radiant tranquility~~
~~Oh, Oh what a really nice day it is~~
~~Oh perfect lunch of tea and sandwiches~~
~~Oh luxuriant~~
~~Oh stupendous and awesome~~
~~Oh~~
Oh fuck.

March 25

Took another long walk in the woods today. I got lost.

March 30

If I could only find a little food and water, I'd be ever so grateful. At least I remembered to bring my notebook with me. Maybe I'll come up with some really good ideas while I'm out here.

April 13

Well, things aren't looking so good, are they? I mean, after all these years, I think I'm finally having to face up to the fact that I may not be cut out for this wilderness stuff. It's not that I don't like the woods, you see – it's just that they all look so much the same it's ridiculous. I've been sleeping underneath this fallen redwood next to the lake for almost a week now, and I'm afraid to go very far because at least I've got water here. I mean, if I go very far, I'm sure to get lost again, and then what? Starvation, exposure, mutilation by bears? Perhaps I'll be boiled in a pot and eaten by wild savages. No, I may not be happy sleeping underneath this tree next to the lake, but I can think of a lot of worse places I *could* be.

April 17

Whooh. Never thought I'd do it, but I finally found my way back, and all by myself without a compass or anything. What I did was wait patiently there at the lake until Miss Blandishment and her students from Anne Bradstreet Grammar came by on one of their field trips, and then I walked back with them. I must admit, they didn't look that happy to have me along, since I haven't bathed in quite a while (the lake is too cold) and I haven't shaved either. I'm beginning to smell a bit gamey, too, I suspect. So they showed me the way into town, but from there I found my way back to the cabin all on my own.

The cabin was just as I had left it. There were only a few differences, which being the fragile shoots of a vegetable garden beginning to sprout from the front grounds, and inside the cabin sitting next to a smoldering fire and knitting a sweater was an Indian woman I've never seen before who tells me she goes by the name of Wampum.

'Excuse me, Wampum,' I said, about as politely as I could. 'But you see, this is my cabin, and while I'm very grateful to you for having taken care of things while I was off philosophically musing in the woods and all, I think it's probably time you should leave. I mean, I hope I'm not being ungracious or anything.'

Wampum didn't really pay me any mind. Her hands just fidgeted at the pleasantly accumulating lines of wool in her lap. She was not that good looking, but there was something gaunt and angular about her face and eyes which was not unattractive, I must admit.

'Piss off, White Eyes,' she said.

At night she slept in my bed, so I slept on the rug next to the fire.

April 18

'I guess I just don't know sometimes,' I confessed to Wampum today. 'Sometimes I think I know, but then I don't know, really. I think I know who I am and what I want to do, but then when I try to be that person or do those things, I seem to make quite a mess of it, actually.'

Wampum was stirring something in a large brass pot over the fire. Sometimes, if I sat very quietly and didn't get in her way, she might fix some for me, too.

'Like this whole journal idea,' I said, trying not to make the empty plate balanced atop my knees look too obvious. 'What could be simpler, right? I just write down what I feel and think, isn't that it? But every time I sit down and try to write something, it's like I don't know what I think. It's like I don't know what I feel. I don't have any good ideas, Wampum. I'm trying to express myself, but I don't seem to have much of a self to express.'

Wampum sniffed again at the bubbling stew. Then she went to the cupboard and returned with a large tin plate. The plate was dented on one side, where I dropped it once while moving.

'You expect too much from yourself, White Eyes,' Wampum said, and began ladling the meaty stew on to her plate. 'Just try to relax and stop trying to be so brilliant all the time. Think one clear thought, then another. Don't rush yourself. For example, tell me what you're thinking right now. Something normal, even. Whatever's on your mind.'

I thought for a moment. I watched the steamy food being ladled on to Wampum's white plate. Finally I said, 'I love stew because it's filled with fat carrots, potatoes, and greasy rabbit chunks. I like the taste of it in my mouth. I like to wipe the rich gravy off my face and hands.'

Wampum took her stew to the good rocking chair and sat down. She straightened her skirt, then rested the plate on her knees. 'There,' she said. 'That's a good start, isn't it?'

Then she commenced to eat her delicious-looking rabbit stew.

19 April

Later that same night I was lying on my cot as quietly as I could when Wampum came over, climbed under the blankets, and did a number of rather unmentionable but not altogether unpleasant things to my body much too quickly. Then she got up again and went back to her bed.

I lay there slightly dazed for a while. There were bluish embers glowing in the fireplace.

'Wampum,' I said after a while. 'I think I just had an idea. What if you stayed here and lived with me for as long as you wanted? Then it would be sort of like Nature had come to me, and I wouldn't have to go out looking for Nature all the time. I am quite interested in the relationship between Civilized Man and Animal Nature, and you, Wampum – and I certainly hope you don't take this personally or anything – are a sort of animal-like person in many ways. Most of them *good* ways, mind you, but animal ways nonetheless. What if you stayed on here, and I began contemplating the history of civilization and nature and all that, and maybe even I wrote a book or something. Would you like that, Wampum? I think *I'd* like that – if *you* did, that is. Wampum? Are you listening?'

I sat up on my cold thin cot. The bluish embers weakly illuminated Wampum's dark face. After a while, Wampum opened her mouth and took a long, snorey breath. She was missing a few obvious teeth. The room was still rich with the aroma of Wampum's simmering rabbit-stew.

Quietly I got up and went into the kitchen for a spoon.

27 April

Sometimes now I wander aimlessly in the woods for hours. Trees, rocks, clouds, caverns, dappled water, blossoming flowers, green grass and black dirt. If there is a certain sadness in nature, I am beginning to realize it is a sadness which nature cannot live without. The arrangement of black rocks encircling the base of a tree. The knotty bark of that tree, the twisted clouds in the sky. Birds, trees, leaves, flowers. I don't know the scientific names of these things, or even anything useful about them. Yet nature is never really incomprehensible to me, either. Nature picks things up and puts them down again, and where it puts them down, somehow those things seem to belong. If Man picks something up and puts it down again, that thing is not where it belongs at all. That thing just lays there like a contradiction.

'It is almost as if that is the definition of Man's intelligence,' I tell Wampum tonight after dinner. 'Man's perverse ability to damage or ruin Nature, to break the secret continuity of things. Intelligence is not constructive, but destructive. It is not a capacity to know, but an ability to wreck.'

Wampum was sitting at the table playing a game of Solitaire. She seemed puzzled by something, and looked from the cards in her hand to the tableau of cards on the table. Then she looked at the cards in her hand again.

'Do you think Man really belongs in Nature at all, Wampum?' I was cleaning the dishes with a soapy rag. 'Now that I'm beginning to come up with a few halfway decent ideas, I'm finding that a lot of the ideas scare me a little. I see the future as this expanding network of hard concrete roads with monstrous, oily machines roaring in every direction. I see gigantic supply stores where people buy fruit and vegetables from other people they don't even know. Everybody sits in their enormous homes all day surrounded by bizarre electrical devices. Civilization is gradually eliminating Nature from the face of the planet, and people don't seem to mind one little bit. In fact, they not only seem happy, but even quite smug about it.'

Wampum closed her eyes. She was still taking slow, reflective draws from her sizzling pipe. 'What has happened to my people in this world of yours?' she asked.

'Your people, Wampum – and this is the saddest part of all – aren't anywhere to be found. There's a few living on these very shoddy reservations, but most of them have gone completely crazy, or become criminals and drug addicts. The skies are filled with gigantic metal spacecraft. White men journey to the moon and build restaurants there. They journey to other planets. They set off horrible chemical explosions wherever they go, and transform beautiful landscapes into poisonous garbage dumps and vast parking lots to accommodate their various transportation devices. Are you getting scared, Wampum? I think *I'm* getting a little scared. I don't think I want to have any more ideas for a while. Do you think that's possible, Wampum? Do you think it's possible that for the next few weeks or so I could give my overworked brain a little rest?'

Wampum didn't say anything or move a finger. The corn-cob pipe turned cold and smokeless in her hand. The uncompleted card game lay spread out before her on the table like an elaborate code. After a few minutes, she began faintly snoring.

Eventually I lay down on my cot, but I couldn't go to sleep for a long time.

2 May

My mind hasn't stopped working even for a moment, and Wampum says I am driving her totally insane. I pace about the cabin all day getting in Wampum's way when she tries to cook or clean. My gesturing has grown wilder and more uncontrollable, and I always seem to be knocking into things and breaking them. 'White Eyes,' Wampum has told me, again and again, 'I hate to admit it, but I have actually grown rather fond of you over the past few weeks or so. However, if you don't shut up pretty soon, I will have to kill you. I really mean it this time. I know when I said it the last few times I didn't mean it, but this time I really do mean it. I will have to kill you and bury you in the woods. And I don't think that's the sort of eventuality which will make either of us very happy in the long run.'

'Death wouldn't solve anything,' I said, disregarding Wampum's deep, almost fathomless sigh. 'I have already begun exploring the possibility of a future in which death no longer exists. In this future world, all of our souls – even yours and mine, Wampum – have been brought back to earth and implanted into these monstrous-looking machines. Our alien mechanical bodies roam eternally across Earth's ruined landscape, crushing out chemical fires and repairing fundamental services for the only surviving fleshly creatures of our world. These creatures have enormous bald heads filled with fat, disused brain tissue, and they sit around all day eating processed food and being entertained by machines like ourselves battling to the death in this gigantic arena. I cannot tell you how disconcerting it is to me, Wampum, to imagine you and I, our secret faces hidden from one another by the complicated steel appendages of our mechanical bodies, battling one another to the death while these stupid people sit watching us and cheering rather mutedly. No matter how hard I try, Wampum, I can't stop imagining this horrible future. It's like I'm on a train and I can't get off. I can't stop thinking, Wampum. Please help me. I can't sleep, I can't eat, and I *know* I'm driving you crazy, I just can't help myself. I hate my brain, Wampum. I really hate my brain.'

Wampum has been sitting very quietly and, I think, patiently as well beside the extinguished fireplace. The cabin is very cold, but neither of us seems to care. She holds an enormous meat cleaver in her lap. Every few minutes or so, she picks it up and examines it.

'Kill me, Wampum. Kill me and make everything better again. But oh, Wampum, what if there really is life after death, and my spirit doesn't shut up either? What if I continue driving myself crazy with all these endless speculations for eons after I've died, and what if your spirit has to listen to me forever, too? I could make all eternity miserable for you then, Wampum, and it would be so unfair, because you wouldn't deserve it. You've been so nice to me, Wampum. You've fixed me such nice hot meals and everything.'

I couldn't help it. I started to cry, sitting there alone on my clumsy cot.

Wampum replaced the meat cleaver in her lap. A rather sad, unpleasant expression crossed her face.

'You may have a point, White Eyes,' Wampum said. Then she got up, took the meat cleaver to the kitchen, and washed it in the sink. 'This situation may call for less drastic measures.'

3 May

Today Wampum took me to see her friend Dwayne, a large very obvious-looking Indian who lives in a squat on the main road outside town. Dwayne sits there patiently every day underneath his modest lean-to and offers to knock mud off the wheels of passing buggies with a big pointed stick. In exchange, the drivers of the buggies toss him filthy, grime-encrusted coins. Dwayne and I hit it off right away.

'Imagination is man's way of addressing the world,' Dwayne said, offering to share a rather alarming hunk of stewed meat with us from his knotty wooden bowl. 'I mean, there's the vast world, right? And then, on the other hand, there's you. It's not the easiest opposition to master. You can't wake up every morning thinking, well, might as well lie around in bed today and see what the world does to me. Of course not. You've got to pretend to have *some* control over your own life, and that means control over the world around you. That's where imagination comes in. Imagination invents rules, laws, essences, causalities, proverbs, truths. Imagination works very hard to invent a competition between itself and the world that only imagination can master.'

'Isn't that good?' I asked. 'Then we can write books, right? Then we can sing songs.'

Dwayne slurped some of the hot, bloody juice from his bowl. 'You tell me,' he said.

'I think some ideas are good,' I said, looking at Wampum out of the corner of my eye. She was sitting next to Dwayne's fire, meditating. 'And some ideas are bad. What's really starting to drive me and Wampum up the wall, though, is not that my ideas are good or bad, but that I keep having them.'

'Are your ideas of the earth or of the air?'
'Both.'
'Fire or water?'
'Both.'
'Men or animals?'
'Men and animals,' I said. 'Boats and trains. Trees and rocks. Rabbits and clouds.'
'Hm,' Dwayne said. Delicately, he replaced the half-gnawed meat in his bowl. Then, after a little while: 'Are you having any ideas right now?'
'About a million of them.' I considered confiding some of my ideas to Dwayne, but at that moment I noticed Wampum had picked up a large rusty knife from beside Dwayne's fire and was examining it intently. 'I have ideas all the time. The funny thing, you know, is that I used to hardly have any good ideas at all. Now I have more good ideas than I know what to do with. It's like feast or famine. Have you ever heard that expression, Dwayne? It's like feast or famine.'
'I'll ask around,' Dwayne said. 'And see what I can learn.'
'I'd appreciate that, Dwayne. I really would. Oh, and Dwayne?'
'Yes?' Dwayne had reached for his pointed stick. A rather disreputable and untidy-looking cart was rattling towards us, pulled by a disheveled and piebald horse.
'Where'd you get a name like Dwayne, anyway? I mean, what sort of name is Dwayne for an Indian? If you don't mind my asking, that is.'
'I don't know,' Dwayne said, pausing to give me a long critical look up and down. He didn't seem to like what he saw. 'Where'd you get that stupid-looking waistcoat?'
'Never insult an Indian's name,' Wampum told me later as we walked home. 'Insult an Indian's name and you're asking for big trouble. And I'm talking big trouble with a capital B and a capital T.'

4 May

'Maybe Nature doesn't really exist at all, Wampum. Maybe Nature's just this idea we have about ourselves. That we have this other place we can go to that isn't noisy or filled with people. Maybe in reality our world is just this one gigantic city, filled with trolley-cars, concrete footpaths and enormous buildings. Perhaps we are all paying to have our brains hooked up to complicated electrical devices which help us imagine green forests, loud birds, blue water. Nature's this idea we use in order to believe ourselves more free. Even when we're in chains. Even when we don't know where we are to begin with.'

I was lying again on my old, now unfamiliar bed in the empty cabin. The hearth was cold and filled with hard black charcoal blocks. There was no smell of food cooking, only the rancid smell of neglected garbage.

Sometimes at night when I can't sleep, I get up from my cold bed and journey into the colder woods, where I am struck by a relatively new idea which has been occurring with increasing frequency over the long weeks since Wampum left me. Perhaps having ideas about things prevents you from seeing the actual things themselves. I have to think about this one for a while. If ideas are something we form inside our head, what good are they for explaining things outside our head? Perhaps we use these ideas not to understand the world, but to make the world into something we want it to be. Ideas are like tools, guns, buildings, governments. We take them into the wilderness and build things that weren't there before. As usual, by this time I have begun to develop a splitting migraine. I look around me at the nameless plants, trees, flowers, clouds, bushes, leaves. There's got to be a simpler way of figuring all this out. I wonder if Wampum knows that simpler way, and if she took the secret with her wherever she went to get away from me. I am standing beside a tiny puddle which is filled with reflected trees and spinning insects. I can't see my reflection in the puddle, but then I haven't really approached close enough yet to look properly.

Then I hear rustling in the bushes nearby, and when I look up I see an enormous, vicious-looking bear with gigantic sharp claws and pointy white teeth coming straight towards me. I cannot believe it. My life is over at last. I will never have another idea ever again, and this thought fills me with wonderful feelings of peace and sadness.

'Hey there, Stupid-looking Waistcoat,' Dwayne says, pulling off his bear-hat and putting it under one arm. 'How you doing?'

It took me a moment to recognize him. Then I had to re-recognize the bear mask under his arm. I said, 'I wish I could say I feel okay, Dwayne. I really do. But I'm afraid I'm not okay. I'm not really feeling very okay at all.'

'Stop whining, Stupid, and listen up.' Dwayne pulled off his bear-claw gloves, then reached into his large stomach pouch for some tobacco and rolling papers. 'Care for a smoke?' he asked.

'No, thank you,' I said.

'It's like this, Stupid. I've been giving this whole problem of yours a lot of serious contemplation, and I've come to the decision that all these goofy ideas of yours aren't really the issue. I mean, everybody's got goofy ideas. And then, of course, there's the subconscious. There's of course a whole *mess* of goofy ideas in the old subconscious. The problem isn't that you *get* a lot of weird ideas, Stupid. The real problem is you keep *telling* them to everybody. You hear what I'm saying? I mean, like, repression's a pretty good deal sometimes. Some things *need* to be repressed. I think you should go on having whatever goofy ideas you *feel* like having, Stupid, and don't let anybody tell you different. But while you're at it, maybe you could find a hobby or something that will keep you occupied during the evenings, so you don't drive Wampum and all her relatives completely nutty all the time. Comprende, pal?'

'Her relatives?'

'Well, most of them are related. There's a few friends of family, of course. And friends of friends, and so on. Wampum went to pack them all up last week. She should be getting back just about now.' Dwayne pulled on his bear mask again. This time he didn't look quite so menacing.

'People have to learn to live together,' Dwayne said. 'And the more people there are, the more learning they've got to do. Oh, look.' Dwayne pointed. 'Smoke signals – and they're coming from your house, Stupid. They say, "Bring home two pints milk, three dozen eggs, three rabbits, one pig." Looks like you're going to have a Homecoming feast tonight. I better run along and get changed.'

With that, Dwayne lumbered off in his ridiculous costume, and I ran all the way home.

12 June

Dear Tribal Family,

How are you? I am fine.

Things here are very good with all of your grandfathers and grandmuthers and cusins and so on. We have bilt some nice teepees and lean-tos to sleep in, and when it ranes we all come and sleep in the house of Wite Eyes, who doesn't seem to mind very much. Wite Eyes is still a very nervus young man, if you ask my opinion, but he is definatly learning to keep his mouth shut when he has to. Also, more importantly, when he can't keep his mouth shut most of us have learned to go about doing our work and ignoring him pretty much all of the time.

Which is to say there is still some mitey fine acres of land around here sutable for all of you to come live on, if you feel so enclined. Wite Eyes is a very mild-mannered sort of young man, and will probably not give you any trouble whatsoever. In fact, I'm beginning to think he may actually apreciate the company.

Wite Eyes, by the way, is presently sitting by the fire trying to rite words on a peace of paper, which continues to be a very anoying preocupation of his. I try to tell him – Wite Eyes, you are a young man with too much intrest in the names of things. You want to know those names so you can rite them on a peace of paper and hide them away like money or savings bonds in one of your nasty government banks. The names of things, however, are a lot less dependeble than that, being as that the names of things are always changing, which is one thing us Indians have known all our lives, if you wite people would just stop killing and robbing us long enough to listen to us for just one minit. There is no such thing as a tree. There is only the red tree by the river that casts small shadows, or the old tree with bared roots where lives the hedgehog with the cut face. If wite men would stop trying to put names on things all the time and own them, they mite have time to open their eyes and see the real world all around them, which looks quite beutiful enough as it is, actually. Beleeve it or not.

Oh, well. I gess I have gone on a little too long, which is to say I'm a lot like Wite Eyes myself sometimes. I gess the old Indian saying still goes:

Nobody's perfect.

Ciao,
your loving neece of the increesingly more pudgy posterior,
Wampum

PS Oh, and I almost forgot. It's a girl.

THE SECRET LIFE OF HOUSES

Mother was admitted to St Jude's Hospital in early July for what the doctors called 'routine tests.' The following morning she underwent surgery for the removal of her left breast and, a week later, was expected to die sometime in the afternoon when Margaret arrived with her schoolbooks and a copy of Mother's *Racing Form*. The thin blond pastor was standing guard outside Mother's door, waiting to take Margaret to the visitors' lounge where, for more than two hours, he spoke to her in hushed reverential tones about the Lord Jesus. 'Your mother and I have enjoyed many opportunities to talk personally about the Lord Jesus,' he said. 'And I think I can say, without fear of contradiction, that she would like you to know that earthly expiration does not mean the end of life but rather the discovery of a heavenly love far greater than you can possibly imagine.' The pastor's voice was filled with gigantic glass-paneled cathedrals and vast, beatific white light. Swelling organ notes pulsed in your bones and heart, and the soft, contented faces of very spiritual, loving people in nice suits and summer dresses turned to look up at a beautiful white man in white flowing robes. The visitors' lounge was a chipped gray pastel, and included a pair of wobbly wood-frame chairs, a malfunctioning RCA black-and-white portable television and a few tattered issues of *People* magazine. While the pastor talked, Margaret imagined herself living in the glass cathedral with Jesus, like a tiny mouse. Being much smaller than all the lovely people in their beautiful clothes meant Jesus would love her even more, and she would fix him good healthy meals in the little basement kitchen. Beef broth, salads with blue-cheese dressing, high-fiber cereals and fresh-baked brown breads. Some days Jesus came home very tired, not in any mood to go upstairs and comfort his assorted brethren. At these times Margaret would have to speak very firmly and lovingly to him. 'Now I want you to go upstairs right this minute, Jesus, and stop feeling so sorry for yourself. I *know* you're tired, but so are all those nice people who have been waiting upstairs all day to see you, and I'm not going to let you disappoint them. Now get out there and do your best, and know I love you very much or else I wouldn't even bother talking to you like this.'

Every day after the first operation when Margaret arrived at the hospital she found Mother asleep in the immaculate white bed. Mother appeared somewhat collapsed and unfamiliar now, like a favorite pillow with the slipcover removed. Her mouth hung open, revealing dull gold and silver fillings. The bright amber diagnostic display hummed darkly beside her bed like a brain. Margaret would sit in the red Leatherex chair beside the bed and read her civic studies book entitled *Voyages to Discovery: Making Friends Around the World* in which a page 37 illustration depicted a primordial woman with large brown breasts hand-sculpting a blue clay bowl. A small brown baby sat naked beside her, ornamented with bright beads. They both seemed very busy and very content. Somewhere beyond the frame of their picture, Margaret imagined, fierce men battled enormous dinosaurs in order to make the world safe for civilization. Primitive men lacked houses, and had to make do with mud huts. Sometimes Margaret examined the pictures of mud huts and tried to imagine what it would be like without the security or real weight of a house around you. The world would seem very shadowy and dim, she thought. The wind would whistle through everything. Parts of your mud hut would be blown away, or eroded by rain. Life would lack depth, weight and dimension. You couldn't hold it in your hands. You couldn't hide among its deepest rooms.

Every evening after visiting hours Margaret would take the bus home to her own very solid house and cook dinner in the toaster oven. She liked to fix Budget Gourmet dinners, because they included vegetables and starch. She prepared large green salads with french dressing and stored them in plastic Tupperware. She also liked tuna melts, toast with jam, frozen pizza. Then she would clean the toaster oven with a damp sponge, wash her dishes in the cracked porcelain sink and fall asleep on the living-room couch with the television on. The television marked a clean, warm place in the otherwise silent house. Margaret's favorite show was *Dallas*, but her favorite actor was Peter Graves on old reruns of *Mission: Impossible*. Dense, compact shadows filled the unseen rooms with a hard, dark weight, like the cartons of Mother's old college textbooks in the attic, or the massive and disconnected Whirlpool Frigidaire gathering dust in the garage. Every Saturday she would take Mother's checkbook from the kitchen cabinet and pay the latest bills, writing out the checks in her best penmanship, affixing a tidy postage stamp and adjusting the return statement until the corporate address showed through the envelope's clear plastic window. Then, on early Saturday evenings after her shopping, she would deposit the week's incoming checks from Social Security and Blue Cross in the Home Federal Savings night-deposit box. The night-deposit box had a heavy steel door which Margaret would pull back with both hands; the dark steel vault hissed with shadows, like an enormous seashell. There was something about the vast and bristling darkness that Margaret appreciated, something about it that seemed to connect her with the vaster darknesses of her own house.

Margaret enjoyed being alone in her house at night because it was her house and Mother's house, and in some ways Mother was still there. Often, while the living-room television generated its warm noise like a central radiator, Margaret would walk from room to room of the house and take random inventories. The rooms retained a formal simplicity, some fundamental resonance of Mother. Margaret was wary of touching anything, of violating the silent rightness of things. At night the white moonlight fell across the taut coverlet of Mother's bed. The lid of the cracked ersatz-leather jewelry box on the bureau contained an unsprung music box which chimed faintly when you opened it; the box also contained broken costume brooches, unmatched earrings, fractured glass pendants. The chipped plastic bracelets and necklaces had begun to fade and oxidize in places. The enormous television was covered with crumpled white lace doilies, pencil stubs, an open *Racing Form*, in which penciled notations and statistics had begun to fade, and a moldering orange in a painted bowl. The bowl depicted scenes of Florence, where Mother's sister Rita had honeymooned in 1957 just before she died. Mother still had a few faded black-and-white photograhs of Rita and her vacationing husband attached to the frame of the vanity mirror. Their features were colorless in the harsh light, like lunar newlyweds vacationing in a sea of dust. Margaret imagined it must be a world without sound, where everything was soft and drained of gravity. Rita died when she was only twenty-four. 'She never got a chance to see our house,' Mother said 'But I know that if she had seen it, she would have thought it was the best house in the entire world.'

On Mother's bathroom sink and counter various vials of pills were spread about, many of them empty and overturned like toy soldiers. When you opened the medicine cabinet, its mirrored surface spattered with Crest and Listerine, you saw antacids, aspirins and Tylenols, codeine and bromides. Margaret's favorite place, however, was Mother's closet. Sometimes she would open the door and step inside between the two musty aisles crammed with clothing on hangers. Outdated fat, fluffy overcoats, beaded flannel robes, moth-eaten gray sweaters and paisley blouses. A brown and beige Denny's uniform displayed Mother's name on a plastic nametag: ANN. The name seemed at once strange and familiar, like a commercial advertisement. Above the racks of clothing boxes of papers and receipts were stored. Bulging hatboxes and shoeboxes with cracked joints, belted with thick red rubber bands, boxes which had formerly contained blow dryers, standing desk lamps, clock radios. Margaret liked to stand atop a small white ledge and browse among the boxes, moving them about but always remembering to replace them in their appropriate spot. Receipts, letters, keys in tiny envelopes, ropes of dust. Often Margaret and Mother had conducted these late-evening surveys together. 'You'll have to know where everything is,' Mother would say. 'In case anything ever happens to me. Not that anything will ever happen to me.' She showed Margaret where the fuse box in the basement was, the emergency phone numbers in the kitchen. 'Here are your grandmother's photographs, and Rita's wedding album, and a key to the safe-deposit box.' Whenever Mother conducted these impromptu tours of her secret economy she carried a rum and Coke ahead of her like a flashlight. They would turn on all the lights and play records loud on the phonograph. 'Here's your Social Security card and your birth certificate. You know why I'm showing you all this stuff, don't you, baby? Because when and if I die someday, I don't want your Aunt Fergie coming over here and taking off with everything. All of this belongs to you. The day I die your Aunt Fergie will be over here like a shot. She'll be driving a U-Haul trailer. She'll pull everything out of this house she can carry. She'll take the knobs off the oven. She'll take the toliet paper off the roll in the bathroom. The day I die you watch out for your Aunt Fergie.' As Margaret stood alone now in the crowded, dim closet, she could still hear Mother's voice resounding in the close, musty walls, the jumbled boxes and heavy clothing, like the beating of some suppressed heartbeat. It was as if while Margaret's eyes had grown accustomed to the contents of Mother's dark room, her ears had grown accustomed to its voice as well. There was something delicate and infinitely priceless about every object, a certain character and voice which Margaret feared she might accidentally disturb or erase. Alone, Margaret never turned on Mother's bedroom lights; she never played the phonograph. Margaret didn't hear the voice so much as detect it, just as she could detect the weight of halved peaches and syrup in an unlabeled tin can. The voice was Mother's voice and it was not Mother's voice. It was the voice of the house, too. Margaret thought of Mother in the breathing ward, surrounded by unconscious elderly women gazing blankly at the fluorescent ceiling.

Later, with the sounds of the house still resounding in her blood and brain, Margaret would drink coffee and watch the *Tomorrow* show, with Tom Snyder, or perhaps a late movie, thinking about Mother in her white bed. There were times at night when she felt as if everyone in the city were asleep except her and her resonating house. She felt like a monitor, a secret trespasser of sorts. Sometimes she wondered if people passing in the street could detect the phosphorescent glow of her black-and-white television through the venetian blinds; she wondered if anyone knew about her secret life alone in the house.

The nurses began paying Margaret little attentions. Cookies, snacks, preteen paperback novels in which young girls of Margaret's age solved crimes, met romantic chums in foreign countries and overcame personal embarrassments at school. Sometimes she was awarded secondhand clothing, which she occasionally accepted, and dolls, which she always disdained. One of the nurses even arranged to have a television installed in Mother's ward; usually, though, Margaret preferred to do her homework.

'Are you good in school?' the nurses asked.

'I guess so,' Margaret said.

'Do you have a father?'

'He lives in Detroit.'

'Do you live with your aunt? Your grandma?'

Margaret thought for a minute. After a while she said, 'I live with my aunt. My aunt's name is Rita.' Then her eyes would return to her civic studies book. After a few awkward minutes the nurses would depart to other wards, cafeterias, vended cigarettes, and Margaret would feel embarrassed for having told them anything at all, as if she had betrayed the secret life of her house to strangers.

Some afternoons when Margaret arrived at the hospital, there was a flurry of activity around Mother's room. Numerous nurses and orderlies in white uniforms would be going in and out, transporting intricate machines on carts. One of the orderlies held Mother's chart, which was attached to a plyboard and metal clipboard; he told everyone where to go, a stethoscope looped casually around his neck like an ornament. Once Mother herself was hurriedly transported from the room on a gurney by two large black men Margaret had never seen before. Mother lay very quiet with her eyes closed, the white sheet pulled up to her neck, dreaming about the house, the boxes of papers, the safe-deposit key. At these times Margaret would quietly return to the visitors' lounge with the somber, thin-lipped pastor, where she tried to do her homework while he continued describing to her the House of the Lord. The House of the Lord was filled with many rooms and many mansions. There was a room for Jesus, and a room for Mohammed, and a room for Buddha. There were rooms for Catholics, Protestants and Jews. All these rooms were of equal size and fairly distributed, because all men were equal in the House of the Lord. Every once in a while Margaret would look up at the pastor and try to register a serious moment of eye contact, but once she had done her good deed she let her gaze quickly return to the civic studies book which lay open in her lap.

Margaret's house had two bedrooms and two bathrooms, a living room and a kitchen. The enormous downstairs garage, which Margaret infrequently disturbed, filled with Mother's Ford Galaxie automobile, crates of dishes, and a water-stained rolltop desk Mother had inherited from her stepfather when he died. Dust was everywhere, and a gathering puddle of oil near the tailgate of the Ford. There, deep in the house, Margaret realized, her house opened on to deeper rooms and houses too, just like the House of the Lord. Somewhere deep in those rooms Mother was rolling on her gurney through enormous, echoing cement chambers filled with black and Chicano men who smoked cigarettes while doctors tried to poke and prod her with sharp and complex instruments. Underneath still deeper were even vaster discordant parking garages where automobiles raced and honked and surgeons scrubbed their hands at large concrete sinks.

'Should I call your Aunt Rita?'

The pastor had fallen asleep on the stuffed chair. A tattered three-year-old copy of *People* magazine lay overturned on his lap. He snored.

Margaret looked up at the nurse. The nurse's face flashed with harsh fluorescent light. Margaret didn't recognize her.

'That's okay,' Margaret said, and reached for her purse. 'I'll take the bus.'

All weekend Margaret dreamed of Mother deep underneath the foundation of their house, rolling down long hospital corridors. Mother was very young in these dreams and very beautiful. Mother said, 'We live in the house together, Margaret. We need toilet paper, milk and Shredded Wheat.'

When Margaret awoke on the living-room sofa the house was filled with light. Light from the moon, light from the elevated steetlamps. The entire house glowed with light. It didn't seem like night at all, but rather some weird inversion of it. The house seemed really massive now. Vast underground caverns opened on to still vaster, deeper caverns and passageways. Rooted in the deep earth like a tree, the house articulated with and overgrew other roots, the secret passages of other houses. It's moving toward the hospital, Margaret thought. It's reaching into Mother's room. The house was trying to pull Mother back into the world where Margaret lived. This is my house, Margaret thought. This is where I live. Then, after a while, she fell back into her dreams again, less articulate and threatening now, filled with cool, calm voices. Somewhere she seemd to touch on the periphery of other dreams, and realized, drifting to the surface of her own, that somewhere she converged with the dreams of her own house. The house dreamed fantastic dreams of self-fulfillment, overwhelming pride, spontaneous transformations. The house dreamed it was an ocean liner, a vast white iceberg, a right whale boiling with plankton and animalcula, a duplex shopping center with multiple cinemas. The house heaved up on its concrete legs, ripping at the earth, pulling out long, dwindling complexes of nerve and tissue. The house moved. The house walked.

*

The following evening when she returned home Margaret tried to phone her Great-uncle Ralph in Volcano but his line was busy. Every few months or so Uncle Ralph would visit and sit drinking all night with Mother in the living room. They talked about all the people in the family that they hated, and when Uncle Ralph left, Mother would admit that she hated Uncle Ralph, too. 'He's the tightest man I've ever met,' Mother said. 'Have you ever seen him *once* bring his own bottle when he visits? You watch next time. He *never* brings his own bottle.' Margaret wanted to call and tell Uncle Ralph not to visit this month. She wanted to tell him she and Mother would be away. If Uncle Ralph came and found Mother in the hospital, he might tell someone, and that someone might take Margaret away. That someone might take Margaret away from her house, and then everything would be wrong, nothing would ever be right again. She tried calling Uncle Ralph every five minutes or so, but his line was always busy. She even dialed the operator, and asked if she could interrupt. After a long wait, the operator said, 'I'm sorry. The line is not engaged. Would you like me to try another number?' Instead, Margaret took her blanket on to the couch and tried to resist the house's dark undercurrent, the pull of its turning dreams. This is my house, she thought. Mother says this is my house. She thought she heard sounds outside on the front stairs. If they come to take me away, I'll come back. I'll sneak in through the garage window. The noise halted outside; then, after a moment, clumped up the front stairs. It was quiet again. Margaret imagined the head nurse, surrounded by uniformed patrolmen. She was taking pink and white forms out of her big black bag. A pair of enormous handcuffs gleamed in the moonlight.

But if they came she wouldn't let them in. This was her house, she thought steadfastly. Hers and Mother's. 'That's the way it'll stay,' Mother always said. 'If anything happens to me, you'll get the house. Everything goes to you because you're my immediately surviving heir. Like when your grandfather died. That's how inheritance works.'

What Mother had often referred to as Grandfather's 'estate' was in the garage, shrouded and damp. In a way, Margaret figured, that made the house partly Grandfather's, too. Margaret had never met Grandfather, but she had visited his house once. Grandfather's house was filled with antique furniture, delicate white china in a glass and mahogany case, assorted bedclothes and kitchen utensils, and a fox-fur coat with head and arms dangling like weird growths. 'He's not really your grandfather,' Mother said, pushing open the door of the strange apartment and ushering Margaret inside with a large high-beam flashlight. 'He was basically just your first-rate son of a bitch. He died right there on that bed.' The flashlight's beam drifted across a sudden white coverlet, as sudden and improbable as a glimpse of the moon, then slid across a closer wall and into the kitchen. 'Is that a blender?' Mother asked out loud, entering the cold kitchen with a certain reverence. 'That's just what we needed, isn't it? A blender.' The place smelled close and pungent, like black men on a bus.

It was a few days before Christmas, and when they returned home they redistributed Grandfather's belongings throughout their own personal house. 'I'm seeing my lawyer in the morning. The bastard died intestate, which means I've got to file something or other to be declared executor. There's only some cousins on his second wife's side. They'll probably raise a fuss, but fuck them, that's what I say. That bastard was my father, and he didn't even call me once in seventeen years.' They fixed hot toddies in their new blender, then assembled the thinning aluminum Christmas tree Mother had purchased at Walgreen's three years before. The tree shed twisted bits of aluminum; sitting underneath the tree and watching Mother pour the hot toddies, Margaret felt as if they were worshipping some archetypal television antenna. Margaret was awarded a hot toddy as well, then another. The eggnog was tepid and somewhat scummy. The alcohol scalded her throat, and she felt herself growing more distant from the room in which they sat, the long-familiar anecdotes Mother shared again and again, the voice and weight of Grandfather which was silently assembling in their basement. It wasn't a voice which ever said anything. It was a voice which settled underneath all the other voices and seemed to hold them in place, like the dense colored gravel at the bottom of a tropical fish tank, or the hard gray dirt in their neglected garden. Eventually Mother took the entire matter to court for two years and they inherited the disused Whirlpool, the assorted basement furniture draped in spotty sheets, the living room's stuffed chair, the blender, a toaster oven and four or five thousand dollars in assorted bonds and savings accounts. The fox-fur coat hung in their closet now, its beady glass eyes glinting dully like some secret retribution. Margaret had never met Grandfather, had never even seen his photo. Grandfather was with the assorted furniture in the basement, like a collection of items in some charity thrift store, a certain weight and dimension which altered the weight of Mother's house and voice.

Seventeen years of silence and iron disregard. Cartons of stale cigarettes and personalized matches. Retirement memorabilia from Brisbane Brake and Drums. The aroma of black men on the bus. Mother's house was always filled with ghosts.

Then, one Friday evening, Margaret returned home to discover her Great-aunt Fergie in the living room, removing her large, overgrown black wool coat and laying it across the back of Mother's favorite armchair. A battered brown plyboard suitcase, ragged with peeling baggage labels, sat open on the sofa. Aunt Fergie took a drag from her Camel filterless and flicked one long gray ash on to the floor. She stared at Margaret and took one overly appreciative breath. 'Well, so you must be my little grand-niece. You must be Margaret.' Margaret closed the front door, but did not lock it. Far underneath her feet the great house exhaled a long inaudible sigh.

'I took the bus to Los Angeles and had to transfer. I spent three hours in the Los Angeles Greyhound station. I couldn't wait to see you. I was so sorry to hear about your mother.' Aunt Fergie brought a pot of tea and placed it on the living-room coffee table. 'The hospital called me yesterday and I came right here as soon as I could. They couldn't reach your father or your Uncle Ralph, which is certainly no surprise. Your Uncle Ralph is off boozing somewhere. And your father, well.' Aunt Fergie held her teacup under her nose, reflecting on the unraveling coils of steam. Her eyes seemed very far away. She had gray hair with a few stray flecks of black in it, tied up in a bun under a thin, gauzy hat. Her cheekbones were taut and polished with rouge. Margaret felt the entire house growing very cold and lifeless in the presence of Aunt Fergie, as if the electricity had been disconnected, or the fireplace plastered over. 'It's such a terrible shame,' she said. 'Your mother was always so young and full of life.' Aunt Fergie's face was expressionless. She put down her drained teacup. 'I'll sleep in your mother's room tonight, and we'll go to the hospital together in the afternoon. You still have school, don't you? We could always take you out early in the afternoon. I can write you a note.' They sat quietly together on the sofa for a while. The television gazed darkly at them both, registering a murky soundlessness in the dark room. After a while Aunt Fergie crushed out her cigarette in her teacup and her dim, abstract eyes focused on the large RCA console.

'Is that color?' Aunt Fergie asked after a while. 'Or is it just black-and-white?'

It wasn't Margaret's house anymore. At night she slept in her cold, voiceless bedroom, gazing abstractedly at the posters of Madonna, Mick Jagger and Prince which decorated her walls. The tiny shelf of preteen paperbacks, the disused dolls and board games on the closet shelf. A large stuffed bear lay beside her on her large pink coverlet like some transient boarder. None of the things seemed at all familiar; everything seemed vaguely impossible and irreproachable, even the shadows which fell across things. The house was cold and silent, and when Margaret slept she dreamed of dark abstract spaces filled with strange people. She remembered her mother in these dreams, but she could not hear her voice. The house was growing very cold and formal, retracting from the dark earth, its great bronchial passages collapsing and decomposing into long, shredded strips of black, inorganic matter.

'It's important to prepare you for the worst, dear,' Aunt Fergie said, sitting beside Margaret on the bus and leafing through a Montgomery Ward catalog. 'Your mother's lungs have collapsed, and she's only being kept alive by machines. She's been completely unconscious for more than ten days, so there's no way the doctors can assess what sort of damage has affected her brain. She has indications of spinal meningitis. She could die any day.' Aunt Fergie turned to the Housewares and Appliances section and grew silent for a while. Outside the bright sun flashed on the pale sidewalks. Every once in a while the bus hit large potholes and the entire bus lurched, Margaret's seat rattling. Aunt Fergie's long gray finger tapped at a Whirlpool photograph which carried a retail price of $499.99.

'It's a lot like what happened to your Great-uncle Havelock in Montreal. He suffered massive, irreversible brain damage.' Absently Aunt Fergie gazed out the window at the gray, cloudy Pacific. After a while her gray hand touched Margaret's book. 'Don't you feel very sorry for your Great-uncle Havelock?' she asked.

Margaret felt internally displaced, as if even her vital bodily organs were being dispossessed. Every afternoon she returned home from school to find Aunt Fergie conducting secret inventories, peering in kitchen cabinets, fiddling at locked drawers with paper clips. There was something clinical about Aunt Fergie's attentions, as if she were flirting with a lover, or investigating and cataloging the contents of some comatose body with microscopic cameras, like a program on educational television. 'This isn't bad,' Aunt Fergie said, picking at the frayed antique chair Mother had inherited from Grandfather. 'We could always have it reupholstered.' Aunt Fergie never permitted more than one house light on at a time. They shared spare, comfortless little meals of macaroni and cheese, canned soups and cheese sandwiches in the darkening kitchen while Aunt Fergie gazed absently at the Authentic Hand-Painted Seascape Mother had acquired three years before with six dense books of Blue Chip Stamps. Aunt Fergie lived in the house now, Margaret realized. Now Aunt Fergie dreamed of the deep earth, the gravid blue sea, the moons of distant planets. Aunt Fergie blinked myopically at the seascape, chewing her tepid macaroni, while Margaret merely examined her fork on her plate. Her schoolbooks lay stacked and unopen on the kitchen counter. Unread and unattended assignments marked their random pages. School seemed a very distant and uneventful place now, echoing with dull, muted voices and strange, unstaring faces. Margaret spent hours there every day, sitting alone at recess and gazing blankly at the pages of her textbooks, trying to find the words she once read and awaiting the silent evenings she would spend silently breathing in Aunt Fergie's distant and misshapen home.

Sometimes she returned to the house and found unfamiliar men there. An immaculate couple from Century 21 Real Estate, wearing crisp sensible suits. The woman's lips and nails were bright and polished, the man carried a tan briefcase. They both smiled like commercial product spokesmen, and helped Aunt Fergie formulate compliments. 'Look at all that cabinet space,' they said in the kitchen. 'Manageable yard. Two-car garage. Convenient wiring.' Vials and prescriptions were gradually cleared from Mother's bureau, and Aunt Fergie's cold creams and plastic tissue dispenser replaced them. The files and shoeboxes had descended from the closet, accumulating and reassembling in tidy stacks on the vanity table. On a Friday Margaret was introduced to a lawyer who said, 'I'm going to help your aunt organize your mother's affairs,' before Margaret was dispatched to her thin, shadowy room. The lawyer was short and corpulent, and wore a casual Op T-shirt.

One day Margaret found a bulging shoebox filled with letters from

Father on her bed. 'I'm sure your mother would want you to have them,' Aunt Fergie said, brushing the dust from her hands and sleeves. She was wearing black. 'I found them in your mother's closet.' Margaret knew exactly the place, the place where the box belonged, where it had sat accumulating Father's letters for years and years now. The letters were all uniformly typewritten on white ruled stationery and folded into dense packets which often included fragile and yellowing articles clipped from *The Detroit Sun Times* and *Newsweek*. Sometimes the letters also included photostatic reproductions of box seat tickets to the Detroit Tigers baseball game or even the Ice Follies. Usually the letters said,

Dear Ann,
You must be filled with hate and anger all the time. You have kept my daughter from my love all these years because you in your infinite wisdom believe I am not an 'appropriate' role model for her. I pity you all the time for all your hate and rage and for your inability to respect yourself which only makes you hate me more for being happy with who I am. Right now the entire world is poised on the brink of nuclear catastrophe because of people exactly like you, as I hope the enclosed newspaper article clearly proves. I have decided (very FIRMLY this time) to take immediate legal action, and have even procured the legal expertise of Reginald Dwyer and Associates who say my case is airtight and I should have my daughter back any day now. Then you will be unable to poison her with hate and lies about me or my life, and you will be totally alone and deserve everything that happens to you. You will die filled with your own hate and rage and your daughter won't even love you anymore because by then she will be with her father who only truly loves her and doesn't just use her as a weapon against me. I have enclosed my lawyer's business card just so you can see that I mean very serious business.

Yours sincerely,
Jeremy Andrews

PS Legally you can't do anything about it, because my lawyers know exactly what they are doing, and will allow you no alternatives to my personal wishes in this matter.

Usually a brief crumbling newspaper article was enclosed. 'Soviets Purchase Japanese Transistors,' or 'New Trade Agreements with Spain.' Two or three brief paragraphs described rudimentary contractual negotiations, names of key participants. Unnamed State Department officials usually declined comment. Father always highlighted these formal disavowals with a yellow Magic Marker.

Margaret sat cross-legged on her bed and looked at the bulging shoebox which remained untouched at her feet. Eventually Aunt Fergie reappeared at her bedroom door, blinking at the dark moonlit room in her dark wool nightgown and carrying a large gold and ivory hairbrush. She didn't have her glasses on; her face seemed very soft and ageless now. 'I never disagreed with your mother about how to raise you, but I do disagree with her attitude toward your father. Maybe she doesn't like him, and maybe *I* don't like him. But he is your father, and it's perfectly all right that you should love and get a chance to know him.'

After Aunt Fergie closed the bedroom door Margaret sat quietly for a while. 'I always knew where the letters were,' she said out loud, to nobody in particular. 'I always knew where to find them.'

At the hospital the large amber monitors seemed to grow noisier and more eventful as Mother's body grew less and less, as if they were not simply monitoring Mother's life forces but actually translating them into mechanical languages. Margaret couldn't find Mother in the hospital at all anymore, and simply sat beside Aunt Fergie and her knitting, gazing up at the flickering television, straining to hear its dim, humming voices. Aunt Fergie always kept the volume turned way down. 'So it won't disturb your mother,' she would say, taking up her ceaseless knitting again, blinking behind her thick glasses as if she were sending back telegraphic signals to the fleets of advisers she had already begun mobilizing around the house, like the toy soldiers deployed by boys around beachheads of sidewalk and garden.

The only times Margaret could still find Mother was when she descended into the cool, weightless garage of their trembling house at night like a deep-sea diver. While Aunt Fergie steadily snored in Mother's transgressed bedroom, Margaret would pull shut the kitchen door, careful of her feet on the dark stairs. Watery moonlight spilled weakly through the single basement window, occluded by overgrown ferns and pale cobwebs. Sometimes she had to stand very still for a long time before she felt anything, that deep involuntary tug beneath the earth. The turn of some buried serpent, the darting of some hurried mole or gopher. Then, after a long while, she might hear Mother's voice, distantly at first, tapping underneath the hard, compact earth like a finger. 'The bonds are in a safe-deposit box at Crocker Bank in Burlingame,' Mother's voice said. 'You take the number seven bus to El Camino Real. Then you transfer to the thirty-four. Now, where are the bonds?'

'At Crocker Bank on El Camino Real.'

'And how do you get there?'

'I take the number seven. Then I transfer to the thirty-four.'

Aunt Fergie disassembled everything. The furniture, the files, Mother, Grandfather, Margaret, Dad. 'It's time we faced this situation properly, like mature women,' Aunt Fergie said, scouting through Margaret's drawers and closets. 'We have to keep a log. We have to get everything organized so that if and when your mother does die, God forbid, everything can get through the courts as quickly as possible. If your mother dies intestate, we have to set up a trust and a trustee.' Aunt Fergie moved Margaret's embroidered footstool into Margaret's closet and stood on it. Her hard fingers, adorned with dull gold rings and bracelets, gripped at the upper shelf. On her tiptoes, she reached out one abrupt hand. 'There's some bonds of your grandfather's,' Aunt Fergie said. 'Some rings and jewelry of your grandmother's.' Blinking, Aunt Fergie pulled a toaster-size box from the shelf and fumbled at the lid. She stumbled slightly on the wobbly stool, her enormous black coat flapping like enormous black wings, and with a rush like water, domino tiles and Barbie accessories crashed from the box and clattered onto the floor. As solemn as a mourner, Aunt Fergie stepped down, clutching the collapsed box in her arms. The box said Realistic Radio Alarm Clock. AM/FM dials. Digital read-out.

'Now your mother is too smart a woman just to misplace them,' Aunt Fergie said. She wasn't looking at Margaret, who stood at the bedroom door holding her foamy toothbrush, but rather over Margaret's shoulder at the open door of Mother's bedroom. 'I've checked with her local bank. They've got to be somewhere in the house. Unless she's got another account somewhere.' They stood quietly together in the doorway. Aunt Fergie's chin grew lax and wrinkled. In the house's briefly resumed silence, Margaret heard the old house flickering briefly like a restored pulse. Then, absently gazing, Aunt Fergie pushed past Margaret and into the reassembled shadows and voices of Mother's diminishing room. Late into the night, Margaret heard boxes being reshuffled and banged about, the audible swirling of generations of inherited and relocated dust like cinematic fog in an old black-and-white movie.

Then, alone in her dark bed in her strange and voiceless bedroom, Margaret would open her civic studies text and remove the tiny safe-deposit key taped inside the concluding chapter. The chapter included many photos of the world's children playing together in crowded Bombay, a New York discotheque, a British pub. When Margaret knew for certain her great-aunt was asleep, she would slip quietly into the basement and gently mollify the buried life of her house.

'The surgery hasn't healed,' Aunt Fergie said, knitting beside Margaret in the room where Mother's impostor lay. Mother's impostor didn't even remotely resemble Mother anymore. Gradually she was beginning to resemble Aunt Fergie herself, with her pale, slack skin and long teeth, her thinning hair and gaunt, expressionless face. 'Arteries near the base of her brain keep opening. Her immune system won't respond, her blood isn't coagulating. They've done all they can do.' Aunt Fergie's needles clacked faintly like tiny teeth. 'I've managed to contact your father in Detroit. My lawyer and his lawyer see no reason why the courts won't allow him, under these conditions, to resume custody.'

Margaret envisioned Detroit as vast white pavements and reflecting steel skyscrapers. There were no houses there, only Father at his typewriter, corresponding with the world, sifting through newspapers and magazines. 'He's involved in some kind of project at work,' Aunt Fergie said, handing Margaret the envelope, 'so he can't come for you personally. But he's sent a ticket. We'll start getting your things together and you'll leave next week.' Aunt Fergie left Margaret at the darkening kitchen table. After a few moments, Margaret heard her own door opening down the hall, drawers opening and closing. She reached for the envelope before her on the dim tablecloth littered with crumbs. The envelope contained a Greyhound bus ticket, San Francisco to Detroit. One-way. Father's letter said, 'I can hardly wait to see you. All the love I have in the world. Regaining the family we've both lost. Sino-Soviet relations effectively disenabling the SALT II accord.' It was signed, 'Your Loving Father, Jeremy Andrews.'

That Friday Margaret took the bus to Crocker Bank in Burlingame and presented her key to a young blond woman. The woman requested two pieces of identification and, after conferring with a man in glasses and a dark blue suit who sat behind a MANAGEMENT ACCOUNTS placard, accepted Margaret's Social Security card and birth certificate. The blond woman took Margaret into a clean, unadorned beige room which contained only a thin Formica table, like one of the luncheon tables at school. 'I'll give you, say, five minutes? I'll be right outside.' When Margaret was alone, she opened the steel box with her key and heard something sounding in the back of her mind, as if a key were being inserted in some other door in some vaster and deeper room. This is it, she thought. The bonds were bundled up in yellow six-by-eight-inch envelopes marked URGENT and FIRST CLASS; the rings were contained by tiny plastic bags, the flap of each affixed with a staple, like rare coins in a coin shop. When she held the rings and the bonds in her hands they felt firm and obdurate, as incontrovertible as the deep earth, as eternal as the hard formal structures of the house which had once persisted there. Margaret saw the dark and shrouded furniture in the garage, the cold and ticking automobile, the concrete washbasin covered with dust like the chrysalis of some gigantic insect. Everything was growing stronger already; the hard and dreaming foundation of the house fluttered like an eyelid; something deeper twitched. The rings and bonds were a gift, a sacrifice without blood, like investing parts of your allowance in a private savings account. She would surrender them all in the shadows and privacy of the garage; she would regain the home she had lost. This might be it, Margaret thought. Everything's going to be better again. Margaret was almost home.

Within a few weeks, Mother returned home from the hospital's dark, improbable world like an astronaut from outer space.

'A home is always a house, but a house isn't always a home,' Mother said as she was lifted on a stretcher and inserted through the back doors of the ambulance. Margaret sat beside her in the long chamber, on a small red seat which unfolded from the side wall. 'I'll never forget buying that house just after you were born.' Mother's face was regaining a little of its lost color; her hair remained sparse and dry, however, her body thin and haggard. 'I was working at Macy's downtown. I never thought I could afford a house of my own. Then Dolger built Daly City. It was what they called a "planned community," because it was designed to include everything one could possibly need to live, just like a miniature city. A shopping plaza, restaurants, schools. I wanted you to go to a good school. I didn't want you to attend one of those schools downtown. They don't have any teachers or books, and they're filled with spics and niggers. I don't mean to sound racist or anything, but I had to be at work all day and I wanted you to be safe. I knew you'd be safe here.'

Mother was looking out the side window and away from Margaret as they entered the housing tract. The interminable rows of identical houses covered the hills on either side of the ambulance, each house with its own identical palm tree planted in its own identical green lawn. When they arrived in front of the house and the attendants withdrew Mother from the back, the Filipinos who lived next door were all standing arranged around the eldest brother's cherry '56 Chrysler. There were more than a half-dozen of them, and they all stopped talking and stared as Margaret led the attendants to the front stairs. 'The yard looks really nice,' Mother said. In the sunlight, she looked impossibly old and pale. Curtains were being drawn open up and down the street, and unfamiliar faces appeared at windows. A small boy stood on the sidewalk outside their house and bounced a small red-and-white-striped rubber ball. 'Here we go,' one of the attendants said. Otherwise everyone was very quiet and solemn, as if this were the anticipated funeral which, scheduled in blue ink, must override any minor inconvenience such as restored health. As all the assembled neighbors stared, Margaret felt as if the most secret parts of her life were being systematically violated. She wanted to return alone to the garage; she wanted to sit alone on the floor and think about nothing for a long time, there underneath the concrete sink where she had sat night after night, the bonds and assorted jewelry in her lap, the flickering spiders spinning and drifting around her in the dark. It all seemed like just a dream now, or some frivolous entertainment on TV. Then, suddenly, it was all over. The attendants deposited Mother in the bedroom and left. 'Have a nice day,' the talkative one said.

After Margaret brought Mother a rum and Coke, Mother held the glass in her blanketed lap and gazed around the room for a little while.

Everything looked very different now. All of the stray newspapers and magazines had been cleared from the floors, and the photographs of Rita and Grandma detached from the cabinet mirror. The broken jewelry box and hand-painted porcelain bowl had been boxed up, along with a few other miscellaneous items of clothing, and stored in the basement. Almost the entire contents of Mother's closet had been removed as well, and the few remaining boxes of papers sat about lidless on chairs and countertops. 'This was always our house,' Mother said. She had not touched the glass in her lap. 'You were raised here. We were the first owners. Even Bill and Julio, they never really lived here. It was always a space just for us.' Mother's eyes were moist, her throat hoarse. 'I paid for this fucking house. I paid for this fucking house when I didn't have it, when all your father ever sent me in the mail was more fucking misery every day.' Mother wiped her eyes with the back of her hand. 'Then I get sick for just *two minutes* . . .' Suddenly, Mother began to cry, and Margaret sat beside her on the bed, gazing out the window at dead, disregarded marigolds in the window box. Mother cried silently, taking long sudden gasps of breath every few moments. The air around Margaret seemed to grow moist; she felt Mother's damp hand grasp her arm. 'I think I'm going to have Aunt Fergie exterminated,' Mother said, and took a long noisy sniff. 'I think I'll have it scheduled for next Christmas. We'll treat ourselves to something really special.' Mother removed her hand from Margaret's arm and used it to raise her glass.

They sat together in the diminished bedroom, the memory of the room's former litter circulating vaguely in their minds like a private atmosphere. It was the atmosphere of the house's buried rooms and passages, Margaret thought, the hidden secrets of the worlds we've stored inside. Strange desultory clouds were dreamed down there, orbiting buried tarns and murky rivers. This landscape the house dreamed bubbled and transformed itself, like gestating planets, like evolution, like dreams themselves. 'Tomorrow I want you to take a walk in the yard,' Mother said. Gazing out the window at the darkening sky, Margaret tried to remember the yard but couldn't. It seemed as if she hadn't been there for years. The grass was certainly brown and dead by now, the roses and nasturtiums. Margaret wanted to tell Mother that Aunt Fergie's voice invested the house now too. I'm sorry, she wanted to say. It's not my fault. She could hear Aunt Fergie's voice scuttering about downstairs on the cold concrete, scratching at cardboard, sniffing at stray unidentifiable droppings.

The previous afternoon Aunt Fergie had taken the Greyhound to Volcano where, Margaret learned later, Uncle Ralph had died of a coronary occlusion and had lain alone and undiscovered for almost two weeks in his bright, airy country house with its white lace draperies and polished hardwood floors.

After a little while, Mother began to snore, and Margaret took the half-full glass from her lap. Now, asleep in the gray moonlight, Mother again resembled that impostor in the hospital. Her slack mouth and doubled chin, her large silver fillings and gold crowns. 'I feel so changed,' Mother had said that day, that first remarkable afternoon she emerged from her coma. Nurses with strange wary expressions peeked in at the door, and dazed withering patients in loose plaid robes. 'My body feels so slow, so different. I don't even feel like the same person anymore.'

Margaret left the door partially open, took Mother's glass down the hall and washed it in the kitchen sink. When she shut off the water, the plumbing gave a little tug deep underneath the house, and Margaret wondered when it would all stop, when it would all grow solid and permanent, like a painting on a wall. Turning, reaching, heaving. The house had its own secret dreams and dimensions now, no matter what Margaret or Mother thought about it, its own plans about where it wanted to go and who it wanted to be. Margaret grasped her cold knobby shoulders, felt her own deep eternal body changing too. Buzzing tissues and organs, membranes and arteries, hormones and blood. There was nothing you could do to change or influence some things. Everything lived its own life down there in the basements of houses and bodies.

When Margaret took her schoolbooks into the living room she didn't turn on the lights or the television. She preferred to sit alone in the glistering darkness for a few hours, just listening, and eventually fell asleep on her couch and dreamed the vast dreams of her house.

THE PARAKEET AND THE CAT

'Being yourself is never a very easy row to hoe,' Sid said, nibbling sourly at a bit of unidentifiable root or mulch. 'Being the only stray parakeet in a drab world filled with cackling hive-minded pigeons, sparrows, black crows and pheromone-splashed finches can be a pretty dismal experience indeed. Especially if you're all alone. Especially if you're at all like me, and inclined to be pretty morose at the drop of a hat *any*way.'

Without a doubt, winter was the hardest season. In winter, even the leaves abandoned you, while all the anxious birds you were just getting to know on the high wires departed precipitously for warm, ancestral climes – places with mythic, irreproachable names like Capistrano, Sezchuan, Sao Paulo, Bengal. Other birds never told you where they were going, how long it took to get there, or even invited you along for the ride. They seemed to think that if *they* had a perfectly nice location picked out for the off-season, *every*body did.

'It's sort of like being dressed up with no place to go,' Sid told the ducks at his local pond. 'I mean, my hormones have shifted into surperdrive. My blood's beating with procreation and heat. And here I not only can't find any girl parakeets, I can't even find the goddamn continent of Australia.' The ducks were an addled, pudgy lot, filled with ambitionless quacks and broad steamy flatulence, fattened by white breads and popcorn dispensed by local children and senior citizens. They sat and jiggled their rumps in sparse blue patches of the partially frozen pond. Snow was everywhere. Winter was pretty indisputable now.

'God*damn* it's cold,' the ducks honked and chattered. 'Jesus fucking Christ it's *cold* cold cold. We're freezing our collective little butts off out here.'

Ducks might be self-involved, Sid thought, but they were still a lot better than no company at all. If you kept close to ducks, you might pick up stray bits of cracker or information now and again, or be alerted by the emergency squawks that declaimed wolf-peer or weasel-crouch. The winter world was a hazardous and forlorn place, and you could use all the friends you could get out here. Buses and planes and whirling frisbees and crackling power cables, children with rocks and slingshots and B-B guns and pocketknives. The wide world was filled with angular objects that were always rushing towards you without any regard for personal space or decorum. Not to mention the ruder angularities of solitude and exile. Not to mention hunger, sadness, constitutional ennui, or even just the bloody weather.

Sid had possessed a home of his own once. A tidy gilded cage with newsprinted linoleum, plastic ladders, bells on wires, bright sexy mirrors and occasional leafy treats of damp lettuce and hard, biteable carrot. Beyond the cage, as ominous as history, stretched a lofty universe of massive walls and furniture and convergent ceilings inhabited by gargantuan creatures with glistening forests of hair and long fat fingers. These gargantuan creatures were always poking these fat fingers at you, and making you sit on them. They whistled and kissed, or made tisky chittering noises, as if *they* were really the parakeets and you too damn stupid to tell otherwise.

'You think you're Mister Big Shot when you've got it all,' Sid told the ducks, who were nibbling a waterproof, oily substance into their feathers with irritable little huffs and snaps. 'Food, water, a warm place to go to the bathroom – that's all you care about. You happily climb the plastic ladder, or happily ping the bell, or happily chat yourself up in the flashing happy mirror all day long and everybody's happy, because that's what you're sup*posed* to do, that's the life you're sup*posed* to live. Happy happy happy days, all the happy goddamn day long, happy little morons just pissing their happy lives away in some stupid cage. I mean, you *think* you're living your own life and all, but you're not *really*. You're just living the life that's expected from you. That's why you're in a cage, after all. That's why you're getting all that free food.'

Sometimes, when Sid was feeling especially bitter and over-reflective, he paced back and forth on a sturdy branch that overlooked the most duck-populous rim of pond. He took quick bites out of the twigs and leaves and flung them hastily over one shoulder, like forsaken illusions. 'Then one day you get sick, just a little head-cold, and *you're* not worried. But the gargantuan people outside aren't so optimistic, and want to know what happens to you *then*? Want to know where this happy little free ride takes you for a happy spin *now*? Into the trash can, that's where. Right into the smelly old happy trash basket, whether you like it or not.'

When Sid stood at the farthest, thinnest extreme of branch, he could feel the faint vegetable pulse of the tree between his toes, he could see the widest horizons of frosty blue pond and white winter sky. 'Boy I'll remember *that* day as long as I live. I'm totally headachy and miserable. I'm coughing up phlegm like it's going out of style. I'm feeling too weak to keep my perch, see, so I drop down to the bottom of the cage for a few succinct winks. I guess while I was sleeping, the gargantuan people gave the cage a few exploratory thumps – I think I remember that much. But I was too tired to react. I just wanted to take this long nap at the bottom of the cage, and the very next thing I know, *bingo!* I'm digging my way out of a newspaper-padded shoe box in the trash can outdoors. I'm sick as a dog. I'm so pissed off I can't see straight. I mean, suddenly there's all this *space* everywhere, loads and loads of *space*, fat and white and dense with dimensions I've never noticed before. The always. The indescribable. The *everything*. I just wandered and wandered, and before long I stumble across you guys, and this nice blue pond you've got for yourselves. The rest, as they say, is history.'

Though the ducks might sit and listen to Sid for a while, absently preening one another and snapping at fleas, eventually they grew impatient and irascible. They got up and waddled about self-importantly.

'He's *such* a bore,' the ducks complained loudly, flapping in Sid's general direction. 'Talk talk talk – *that's* all he does. And want to know how often the subject of ducks comes up? Zero times, that's how many. In fact, we're beginning to suspect this guy doesn't have any interesting duck stories to tell *whatsoever*!'

Being as the odds against his ever running across a female of his own species was something like thirteen trillion to one, spring didn't offer much promise for Sid. This made winters especially hard, especially while the snow fell and icy winds knocked you about. Spring was the sort of dream you had to dream *with* somebody, and Sid was beginning to feel this particular dream was one he would never successfully dream at all.

'Have you even *seen* any girl parakeets?' Sid asked whenever he encountered random sparrows or blackbirds. Usually they were strays who had suffered recent illnesses or injuries, their eyes clarified by wild, dispirited memories of flocks that had long abandoned them. 'I mean, they don't even have to be that cute or anything – I don't mind. And if you haven't seen any girl parakeets, how about Australia? It's like this really big vast yellow place – a continent-sized island, in fact. I don't think it's the sort of place you'd ever forget about once you'd seen it. *I* never saw it, and I dream about it every night.'

Sid's dreams of Australia were filled with bounding orange prehistoric-looking creatures and black, canny aborigines who hurled hatcheted boomerangs and fired poison darts from blowguns. Brilliant clouds of parakeets swarmed in the bright sky, crying out Sid's name, wheeling and singing songs about an ever-imminent spring which would surely last forever. Whenever Sid awoke on his cold branch he could still hear those distant parakeets singing. Then he took one glance around the frigid, lens-like pond and he knew. It was only the routine squawking of ducks. Ducks and ducks and ducks of them.

'Look at Screwy!' they cried (for the ducks were eternally ragging at one another, like old maiden sisters). 'He's trying to eat another gum wrapper!'

'And look – here comes Big Bubba Duck. And boy, does he ever look pissed off at Harriet!'

Raging with secret, genetic industry, Sid spent entire days chewing things up. Branches and leaves and nuts, punching through their knotty, fibrous tissues with his sharp hooked beak.

'This is what I do,' Sid told the ducks, flinging bits of wood and pulp everywhere like a miniature buzz-saw. 'I'm a wood-borer. I bore wood.' When Sid wasn't talking to the ducks, he was tearing everything within reach into splintery little pieces. Day after day, hour after hour, sometimes even late into the night when he couldn't sleep.

'I'm a wood-borer,' Sid proclaimed edgily, his eyes wide with something like panic. 'Wood boring just happens to be one of those things I do really, really well.'

*

One day, after a particularly dense snowfall, a cat arrived at the pond, bringing with it a murky, hematic odor of cynicism and unease.

'Hey there, you guys,' the cat said, maintaining a polite distance. The cat was gray, and sat itself smugly on a large gray rock. 'Boy, are you ever an attractive-looking bunch of ducks! Seriously, I'm really impressed. I never even suspected ducks *came* as good looking as you guys, or halfway near as intelligent, either. I guess that just goes to show me, doesn't it? I guess that just goes to show that I don't know that much about ducks after all.'

At first, the ducks glided off warily into the cold trembling pond, pretending not to be bothered, but never taking their eyes off the cat for one moment, either.

'I'll tell you something, guys,' the cat continued, in a voice as gentle and intrepid as desire. 'I just came from the city and you don't have any idea how lucky you've got it out here. What a nightmare. What a cesspool of smog and urine and crime and poverty they've constructed for themselves in the city, boy. Dog eat dog, cat eat cat, cars running *every*body over without so much as a hi or a how-de-do. Bang crash roar crash bang – I've had enough city life to last me a few thousand centuries or so. Which brings me, of course, to why I've decided to move out here to the woods with you guys. Fresh air, sunshine, plenty of exercise. And of course a *strictly* monitored vegetarian diet from now on. I'm taking charge of my life, boy, and taking it on the road. Call me an outlaw, if you wish; call me a rebel. But I'm tired of living the life society *tells* me to live. I'm finally going to live *my* life for *my* self, thank you very much. Come hell or high water.'

While his smooth voice wetly purred, the cat licked his stubby, retractile paws and groomed his long twitchy whiskers, as if dressing himself for church. Then, giving the ducks a last fond look over his shoulder, he rested his head on the large gray rock and fell indefensibly asleep.

'Frankly, I don't think you ducks are exactly the brightest flock of fowl I've ever come across in my rude travels,' Sid said, perched high atop a buoyant willow. 'We're talking a fat gray cat now, and that means cat with a capital C, A, T, and I can't believe I'm having to actually spell it *out* to you guys. Cats are what you call notoriously fond of fowl, fowl being you ducks and me both. We're like this cat's dream of a main meal, and I don't care what he says about wildlife solidarity, or karma, or pantheism, or even free will. That cat wants to eat us alive. He wants to chew our flesh and rip our blood vessels into stringy pasta. But he wants to play with us first. He wants to tease us and cut us and watch us die slow. That's because *he's* a cat, and *we're* what you call fowl. Am I going too fast for you guys or what?'

With the arrival of the cat's sedulous gray voice, a cloud of drift and complacency began to descend over the tiny duck pond. The ducks took longer naps on the bank, and didn't squawk so much, or flap, or flirt, or battle. They wandered off aimlessly into the high reeds and bushes, snapping up bits of worm and seed, cuddling with their ducklings and gazing up at the slow riot of white, hypnotic clouds and mist. It was as if everyone had suddenly ceased dreaming all at once, Sid thought. It was as if expectation didn't persist here anymore, or incandescence, or passion, or blood. Across the pond's cold glaring logic, the only heat was the large gray cat's heat, the only voice was the large gray cat's voice, the only burn and lungy whisper and modular red pulse was the cat's, the cat's, the cat's, the cat's.

'*Carpe diem*,' the cat said. 'Seize the day, live for the moment, enjoy it all while you still can. As long as you've got your health, you've got *everything*.' When the cat wasn't sleeping, he was speaking his low voice across the pond, a voice which the steely surface of water seemed to reflect and amplify, like light or temperature. 'Sleep and eat and make love and party, party, party till the cows come home. Why live for tomorrow when tomorrow may never come? History, genetics, philosophy, evolution, teleology and math – that's the world of the city, pals. That's the world of machinery, concrete, hypermarts, petroleum and death. We, on the other hand, are in and of nature. We make our own rules, define our own characters and attitudes and laws. You may be ducks, and I may be a cat, but that doesn't mean we can't also be really good friends. I'm actually a very thoughtful and sincere individual – once you get to know me, that is. Everybody likes me, everybody trusts me. Everybody perhaps with the exception of my little pal, the parakeet. Isn't that right?' the cat said, peering up into the acute angles of sunlight that intersected the jostling willow like spiritual traffic. 'Isn't that right, little pal?'

That's right, Sid thought firmly to himself, refusing to allow the cat even a glance or a whisper. He didn't want to grant the cat any responses that could be woven into luminous spells of innuendo, gossip, and misdirection. He didn't want his best thoughts and intentions to be mistranslated into cat-like purposes.

And just to clarify the matter a little more precisely, Sid thought, I don't happen to trust you one single little bit.

'Did anybody ever tell you that you have really beautiful plumage?' the cat said, and began licking his knobby paws again, his thick tongue snagging every so often against a stray indication of claw.

I know, Sid thought, staring off resolutely at a distant mountain. My plumage is quite exceptionally beautiful indeed.

Just about the time of the first slow thaw, Sid went for a flight around the pond and discovered the partially devoured remains of a duck concealed behind a copse of blueberries. Flies converged there, and an odor of bad meat and disintegration. Initially Sid felt a moment of giddy, electric self-displacement, as if suddenly confronted by his own reflection in some twisted mirror. The sky seemed closer, the wind harder, the ice colder. Then, with an involuntary volt of panic, Sid leapt flying into the brilliant white sky.

'It was Screwy,' Sid told the other ducks back at the pond. 'And just the way I always said it would be. Lungs, kidneys, liver, brains – that cat even chewed the spleen out of him. We got to stick together, now. We got to keep alert, assign guard duties and group leaders. We've got to remain calm and, *what*ever happens, stay the hell out of the high brush. Women – keep your ducklings in line. Guys – sharpen your bills against that sandy rock over there. This is war, this is Jericho, this is the Final Battle. Evil has come to our pond, and it's snoozing away on that big gray rock over there. Evil has come to our pond in the form of a cat, and that cat wants *all* of us for its next breakfast.'

'Evil?' the cat sighed, reaching out with its front paws for a hard, slow stretch. 'Isn't that a bit much, really? Aren't we all part of the same food chain, aren't we all equal in the eyes of Mother Nature? Don't go getting all moralistic on me, Mr Parakeet. Ducks die in this hasty world of ours, and so do cats – that's the law of flesh. And Screwy, if you remember, was not a particularly astute duck. If you stop and think about it for one moment, you might realize that Screwy could have been eaten by just about *any*body. So don't go blaming me because of my species, pal. My species may be feline, but my heart is true, and I only wish all of you – ducks and parakeets alike – nothing but the best health, the longest lives, the happiest dreams. And now, if you don't mind,' the cat said, resting his chin on his paws, 'I think I'll get back to my own happy dreams for a while.'

'Quack quack,' said the addled ducks, milling about in a dispirited feathery batch. 'Screwy is no longer a duck, the cat is no longer a cat. Parakeets and cats disregard ducks altogether, and only talk about abstractions like Evil and Law and Society and War. This is a little too complicated. This is a little too obtuse. Let's do like the cat does, and take ourselves a nice long nap. Let's all take a nap and dream of fat, meaty flies in our mouths, and a world filled with nothing but other happy, happy ducks.'

Sid began keeping a meticulous census on the extinguishing flock. Fifty-seven, fifty-six, fifty-four, forty. Thirty-nine, thirty-seven, thirty-one, twenty-five. He couldn't always locate them after they disappeared, but he could quickly sense the general attitude of cool and unworried disaffection that possessed the pond's addled survivors like some sort of inoculation. The survivors rarely looked at one another anymore, or exhibited any signs of affection. They allowed their ducklings to wander off unprotected into the high brush, they didn't eat as much as usual, they grew thin, spotty and slightly diarrheic. Sometimes they slept, or drifted aimlessly on the thawing pond among wide broken platforms of ice, or just strolled aimlessly in circles on the bank, snapping at indistinguishable pebbles and insects. It was as if the ducks had surrendered to a force far greater than themselves, a force which permitted them to nap, disregard, wander, delude, demagnify and concede. The world's escalating reality was making the ducks more and more conjectural and abstract. Soon the bank and brush surrounding the pond were littered with splintery duck bones, broken duck bills, and moist forlorn puddles of bloody duck feathers.

'They don't listen to a word I say,' Sid complained to himself out loud. 'It's like I'm talking to myself. It's like we don't even speak the same language anymore.'

'They do what nature tells them,' the cat said wisely, sharpening his claws against the base of Sid's willow. 'That's why they're at peace with themselves, that's why they can sleep and nap and rest. They realize that the universe is just this big blazing oven, burning entire planets for fuel, driving into the long black spaces all alone without any proper destination in mind. Ducks aren't smart enough to trouble themselves with things like morality or justice – and that's where they're one up on you and me. Heaven's a dream they'll dream *after* this life, and not before. Why don't you try it yourself, pal? Why don't you just take it easy and go with the flow. We all die, we all suffer, none of our dreams hold. It's not necessarily bad, or evil, or tragic, or sad. Just look around you – the ice sparkles, the bare trees sway. Winter's a pretty beautiful season, if you give it half a chance. As long as you've got a nice thick coat to keep you warm.'

Every night, Sid grew feverish with bad dreams and black reflections. He tried to stay awake and watch the cat, who never seemed to abandon the perimeter of gray rock. He could see the cat's luminous green eyes in the darkness, eyes that watched him while he drifted away into near-deliriums, lulled by the cat's ceaseless gray voice.

'Come down out of the tree,' the cat whispered. 'Come down and take a bath in the pond. It's not so painful once you know, it's not so scary once you're taken by its teeth. It can even be a very sensual experience, or at least so I've heard. You won't be lonely anymore. You won't suffer. You won't live dreams that always disappoint, you won't feel hope that always flees, you won't know love that always lies.'

Sid began to lose his formidable appetite; he stopped trying to rally the ducks into assuming tactical deployments and responsibilities; he even stopped boring wood. He sat in the high willow all day long simply trying to stay awake, drifting into reveries and naps, awakening with a galvanic thrill whenever he smelled the cat in the tree and looked down.

'I just thought I'd come up and visit for a while,' the cat said, attached to the willow's trunk by four alert paws. 'Do you mind if I come up just a little further? Oh, okay. But maybe later? Maybe later in the week?' And then, with a casual glance over his shoulder, the cat retreated backwards down the trunk of the tree again, his long tail twitching, his fat rump writhing with a slow, almost erotic beat.

What *could* you count on in life? Sid wondered. Loneliness, predation, crepuscular scurryings, agony and death and the cold comfort of abstractions? The ducks in the pond continued to dwindle and nap. Twenty-three, nineteen, eighteen, twelve. Winter, Sid thought. Cold and very white.

'What's it all about?' Sid asked himself out loud, half asleep, nearly submerged by his own watery dejection. 'What's it all mean?'

'Nothing,' the cat whispered. 'Or at least nothing that matters.'

'Who really cares?' Sid asked. 'Who's there to hold you in the night, or hear you cry? Who's there to tell you it'll all be better in the morning?'

'Nobody,' the cat whispered. 'Nobody, nowhere.'

'Why's it worth doing, then? Why should I struggle? Why should I even try anymore?'

'You shouldn't,' the cat whispered, out there in the darkness. 'You shouldn't struggle, you shouldn't try. Just come down out of the tree, pal. Come down here with me and *I'll* take care of you. We're similar sorts of people, you and me. We're different from everybody else. *We* belong *together.*'

At night now, Sid could hear the cat boldly taking the ducks and unashamedly eating them. The ducks hardly made any fuss at all anymore, or emitted any sounds. There was just the rushed guttural purr of the cat, and the wet sound of meat in his throat.

Night and annihilation were everywhere. All day long the fat gray cat slept on the large gray rock.

Sid knew he couldn't last; he knew he'd have to surrender eventually.

'First rule: I don't want to be cut,' Sid said. 'I don't want to be bitten, or tortured, or flayed. I want to go with dignity. Then, afterwards, I don't care what happens. That's just my material substance, that's not my *me.* But I know I can't do it alone. I'm too scared. I know I'll chicken out at the last moment. That's why I need *you.*'

'Don't be frightened,' the cat said, gazing placidly at Sid in the high branches. 'Of course I'll do anything *I* can to help.'

They disembarked one morning after a cold rain. The high dark clouds were just beginning to break apart. On the bank, a small remaining band of thin, desultory ducks slept together fitfully in the shadow of the large gray rock, wheezing and dreaming. First, the cat waded alone into the pond and winced.

'The water's pretty cold,' the cat said. 'But I've got a nice thick coat, so I don't mind.'

As the cat began to paddle, Sid leapt weakly from the willow and landed on the cat's fat behind.

'I want to go out there,' Sid said. 'To the island.'

The island contained a few leafless trees and one broken,

abandoned children's fortress cobbled together with planks of wood and old orange crates. Everything seemed misty and uncertain to Sid that morning. He had not slept properly in weeks. He couldn't keep his meals down. He knew what he had to do, but he didn't feel any desperation about it, any sadness or urgency. He simply knew he was too tired to go on living with the way things were. He had to get it over with while he still had the strength.

'Anywhere you'd like,' the cat said agreeably, looking at Sid over his shoulder, his body coiled and alert beneath the water like an assumption.

'I always thought drowning was the best way to go,' Sid said distantly, watching the island approach, counting his reflections in the rippled water. 'I always thought it would be just like falling asleep.'

'I'm sure it's very peaceful,' the cat said. 'I mean, but in the long run, of course, it doesn't really matter *how* you go, does it?' The cat's eyes were sly half-crescents, void and messageless, like signals from outer space. 'It just matters *that* you go, and with as little pain as possible – that's *my* philosophy. I mean, the only people death really affects are the loved ones left behind – and you're not exactly leaving the world very crowded in *those* departments, pal. Or are you?'

'No,' Sid said. 'I guess not.'

'Death's pretty much over-rated, when you get right down to it,' the cat said, his voice expanding across the pond with a curt gray clarity. 'I don't think death's any more inscrutable than life, really. It's not destination, or conclusion, or loss. I think of it more as translation – a reintegration of the furious self into the selfless, eternally mundane process of *living*. Eating, being eaten, and eating somebody else in return. Because that's all that matters, you see. The vast glorious process just grinding on and on and on. Not truth. Not justice. Not morality or law. And certainly not individuals like us, pal. Individuals don't mean very much compared to the eternal burning engines of the night. Space and planetary explosions and famine and floods and madness and war. I guess I'm what you'd have to call a pantheistic sort of cat, being as I believe – oh my.'

They were only a few yards from the brink of the island, and the cat had come to an abrupt halt in the water.

'What's that?'

There was something brittle about the cat's voice now, something tentative and uncatlike.

'What's what?' Sid asked and, with a succint flurry of wings, transferred himself to a gnarl of drifting branch.

The cat was squinting with concentration, his chin partially submerged beneath the lid of water. 'I seem to have one of my feet caught in something.'

'Reach down with your other foot,' Sid said. Sid was feeling dreamy and sad. Nothing broke the corrugated surface of pond except the earnest, conspiratorial figures of parakeet and cat.

'Yes,' the cat said. 'If I just . . . Oh, that's not it. Now I've got my other foot caught, too.'

'They're what the ducks call slipweeds,' Sid said, 'because they never give you the slip. I realize it doesn't make sense calling them slipweeds – but then, go figure ducks. I don't know if you've noticed, but ducks never swim near this island. Ducks keep to the other side of the pond entirely.'

'Oh my,' the cat said. The cat's eyes were suddenly wider. He spat out trickles of water that rilled into his mouth. 'This is a bit dire, isn't it?'

'Maybe for you,' Sid said.

'Listen,' the cat said, 'maybe if you came over here a little closer . . .'

'Fat chance,' Sid said. 'A snowball's chance in hell.'

'You don't like me very much, do you?' The cat was beginning to struggle and kick a little, like a fish on a line. 'I think you're being very unfair. For a cat, you know, I'm actually a pretty nice guy.'

'I'm sure you are,' Sid said, his own voice enveloping him like a dream. He was trying to keep his eyes open. He was trying not to fall asleep. 'I'm sure you're a perfect saint – for a cat, that is.'

'Oh, hell,' the cat said, and then, with a sudden twist and a plash, his round black snout vanished beneath the surface of water like a midnight vision overcome by harsh, irrefutable sunlight.

Sid was too exhausted to fly. He floated on the knobby branch for hours, hearing the secret rhythm of waves, the moist interior warmth of planets, pausing occasionally to take a sweet sip of water, or ponder his own imponderable reflection in the brightly lidded pond.

'Maybe the cat was right about destinations,' Sid thought, drifting in and out of sleep as the branch rocked, rocked him. 'Maybe, in fact, destinations are places we never get to. Love, home, safety, death. Heaven and marriage and family and hell. Maybe they're just notions of permanence we've invented in order to protect ourselves from the general impermanence of life itself. Maybe the universe *is* an oven. Maybe life really *is* without meaning. But maybe, just *maybe*, these aren't reasons to give up life, but only reasons to enjoy and appreciate it more. Maybe we don't ever get anywhere, or find what we want, or know anything utterly. But maybe that means we can stop punishing ourselves so much, too. Just getting up every day and doing the best job we can – maybe *that's* the most we can ever expect from life. Doing the best job we can every day and then being kind to ourselves afterwards.'

By the time Sid's branch reached shore it was late afternoon, and Sid took himself a little bath. He nibbled damp neglected crumbs from the sand, and felt a tiny kernel of strength blossom in his heart and face. Everything about the waning sunlight seemed slightly richer, warmer, bluer and more real than it had that morning. In the shade of the large gray rock, even the frazzled ducks were beginning to stir a little. Sid couldn't help feeling momentarily pleased with himself.

'Hey, cat!' Sid called out over his shoulder, brushing the water from his wings with a little swagger.

As if in response, a few trembling bubbles surfaced from the blue pond.

'Screw you, *pal*!' Sid said, and then, just as suddenly, he realized.

Completely out of the blue, Spring had begun.